INSURANCE
DISTRIBUTION
OF THE FUTURE

An Interactive Guide for Agents And Brokers

ALEXIS CIERRA VAUGHN

Off Course
Media and Events

Copyright

Insurance Distribution of the Future:
An Interactive Guide for Agents and Brokers

Off Course Media Publishing

Off Course Media, LLC

www.offcourse.ai

media@offcourse.ai

Graphics: Nipa Akter

Editor: Dr. VMST

Book Design by HmdPublishing

Author Photo: Mayweather Images

Copyright© 2025, Off Course Media, LLC

All rights reserved Off Course Media, LLC

ISBN: 979-8-9922648-2-1 Hardback
ISBN: 979-8-9922648-1-4 Ebook - EPUB
ISBN: 979-8-9922648-0-7 Paperback

DEDICATION

This book is dedicated to my Papa (Arthur Lee Holt) and my Grandaddy (J.C. Carroll); they both taught me about entrepreneurship, leadership, and even the importance of insurance in their way. I love and miss you both daily.

To my sister in love, Vacirca "Vadie" Vaughn the real author and writer of the family, thank you for inspiring me with your life and love for writing. I wish you were here to help me write this book, but I could still feel you guiding me, everytime I got writer's block. I hate that cancer took you from us so young, you are incredibly missed. Thank you for being an inspiration!

To Mr. Larry Eubanks, my favorite insurance man. I'm so grateful that you were my family's agent.

ACKNOWLEDGMENTS

To my partner in life, love, and business, my incredible Husband, Marvin Vaughn, thank you for believing in me and being my executive coach when needed. With your guidance, I went from being a good leader to a great one. Thank you for everything, my love; I love you forever.

To my boys Masadah and Giovanni, I'm just Mommy. Thank you for giving me a reason to work so smart. Everything that I do is for you two. I chose and fell in love with this industry because it didn't force me to choose between being present and making a living; I'm grateful for you both being my why. Mommy loves you to the moon and more!

Thank you to my fantastic village: Granny (Luretha), Grandma (Annie Ruth), Mama (Lavania), Daddy (Jeffery), and Mom-in-love (Mimosa). You each play an essential role in my life and my family's. Thanks to all of you, we didn't have to choose between chasing our dreams and building a family.

Thank you to my personal board of directors (mentors) throughout the years; you each played a pivotal role in every aspect of my insurance career.

To the book that changed my life forever, "Thinking for a Change" by John C. Maxwell, thank you for writing it; it shaped my entire career and outlook on life. This is what being a "Big-picture Thinker" can get you; I'm a whole Author now!

A special thank you to the great minds who contributed to the wisdom chapter. I sincerely appreciate you all sharing your wisdom with the readers of my first book; it's a true honor.

CONTENTS

INTRODUCTION

The insurance industry has been around for over a century, and it's still evolving in ways that continue to amaze me. I can't help but smile when I hear fellow insurance enthusiasts—yes, we call ourselves insurance nerds—debate the nuances of policies, coverage, and innovation. It's a shared passion that lights me up every single time.

From the beginning of my career, I've wanted to make a difference in the world, one zip code at a time, and insurance has given me the perfect platform to do just that. Over the past 15 years, I've had the privilege of recruiting thousands of agents and brokers, watching their careers blossom and their impact ripple through communities.

At one point, I tried stepping away from the world of insurance distribution. I thought I needed a break, but the truth is, I couldn't stay away. Six months was all it took for me to realize I wasn't ready to leave an industry I loved so deeply. Instead of running, I decided to embrace the changes and grow with them.

What really reignited my excitement was seeing how much innovation was taking place in the industry. Suddenly, insurance wasn't just about policies and claims; it was about technology, data, and reimagining how we serve clients. I knew I had to be part of shaping this future—especially when I saw that so many conversations around innovation seemed to overlook the distribution channel.

Let's be real: when insurtech first entered the scene, the narrative was all about replacing agents and brokers. That didn't sit right with me. Distribution isn't an afterthought—it's the lifeblood of the industry. So, I made it my mission to be at the table, advocating for solutions that would empower agents and brokers, not replace them.

When I entered the insurtech space, I quickly realized that my experience in traditional insurance and distribution gave me a rare perspective. That's when it hit me: This isn't just about me. It's about bridging the gap between people, technology, and opportunity.

I wrote this book because I want to help close the insurance talent gap and introduce people to the full spectrum of opportunities within the industry. Not everyone is cut out to be an agent or broker—and that's okay! There's a place for you in this industry that fits your skills, personality, and passion.

People often contact me asking how to enter insurance for the first time or transition their agency from manual processes to cutting-edge technology. I created this guide to remove some of the mystery and help you conquer your fears about insurance and tech. This isn't just a book—it's a hands-on resource. Grab a pen and a highlighter, and maybe even form a training group at your office as we explore what the future holds for insurance professionals.

This is your journey, and I want to celebrate it with you. As you work through this book, I hope it encourages and empowers you to take the next step in your career. And when you have a success story to share, don't keep it to yourself! Email me at media@offcourse.ai or tag me in a post on LinkedIn—I'd love to hear how this guide helped you grow in the exciting world of insurance distribution. Let's shape the future of insurance together.

CHAPTER ONE

MY INSURANCE STORY-EXPLORING THE PAST WHILE PREPARING FOR THE FUTURE

Before we discuss the future of insurance, we have to go back to the past. This is where we glean from the wisdom of others and truly reconnect with why we are in the insurance industry in the first place. Revisiting this part of the narrative will help you remember your original why (we know it changes as your career ages in the industry) and continue to inspire you to keep going no matter how bleak the future may look. I'll start by telling you how insurance has been interwoven into my destiny at a very young age, and I didn't even know it. My love for insurance started very early; I grew up in the small town of Dothan, AL, and one big circle connected every part of town. I was the firstborn of two teen parents; my Dad joined the military as soon as I was born. I am the oldest of 14 children; 5 grew up in the same house with my mom. We didn't have much, but we always had each other.

My parents came from two loving families: entrepreneurs, musicians, school bus drivers, and medical professionals. Both of my grandparents had the best relationship; I was both of their first grandchild. I spent a lot of time with both sets of my grandparents growing up. When I was around 5 or 6, I was wherever they were. My Granny and Papa would keep me a lot during the week while my mom worked. Every Thursday, right before dinner, I would watch for that red truck to arrive from my Granny and Papa's front window of their dining room. It was the main window that had the best view of the driveway. I wanted to make sure that I didn't miss his arrival. When he appeared pulling into the driveway, I'd yell, " Granny, he's here!" with the enormous excitement a 6-year-old girl could have. She'd open the door for me to run out to meet him. He'd get out of the truck with the biggest smile and say, "Hey, Cierra, if Granny tells me you've been good, I've got a surprise for you."

The insurance man, Mr. Larry Eubanks, brought me my weekly lollipop, which I had eagerly anticipated. My Granny would say, " Cierra, don't eat that until after dinner," She'd turn to Mr. Larry Eubanks and say, "Come on, Larry, are you ready to eat? Your plate is on the table". Every week, Mr. Larry Eubanks would come to our house to collect what they used to call "Debit life" insurance payments. Growing up, all the black people I knew had regular insurance men who would come to their houses to collect life insurance payments weekly.

In my professional opinion, this is a big reason the data from LIMRA's Black Americans: Fueling the expansion of U.S. Life Insurance report states that" 56% of Black Americans own life insurance. The" Life Insurance Fact Sheet by LIMRA also states that "49% of Black Americans still say they need more life insurance". Back then and even today, Black Americans purchase small death benefits to make sure that they have affordable options to cover their burial cost primarily; we haven't necessarily used them to build wealth.

Insurance agents' past practices included not presenting insurance as a huge benefit to build wealth but more so for the protection of this community because they always focused on affordability, which many of our grandparents and parents also concentrated on.

Mr. Larry Eubanks wasn't just the insurance man; he was a part of our family. I learned a lot of things that I used throughout my insurance career from Mr. Larry Eubanks. He taught me to empathize with my clients, listen, and talk less to understand their needs. He also shared the importance of getting to know your clients so they trust you with insurance and beyond. He educated my grandparents on how to best leverage their policies for loans when it came time for college or big family purchases where we needed a bank but didn't have the credit to access one. He truly taught my family the power of their insurance policy.

He was the insurance man for the "Donnie Mae Drive" neighborhood, named after my great-grandmother, who owned all the farmland our family lived on in the country. When there was a holiday, he'd give us free hams and turkeys. During Christmas time one year, we all got a bicycle from his agency office. He would always think of us and how he could help.

In middle school, Mr. Larry Eubanks and his agency manager gave me a scholarship to a weekend leadership camp for girls; he thought it would be great to have it on my college applications one day. I learned so much from that experience. It was my first time staying in a cabin in the woods, just like they did on TV. I will probably age a bit, but it was like the old Nickelodeon show "Camp Anawanna from Salute Your Shorts." We had a camp counselor and everything.

My Granny and Papa had a majority commercial and residential cleaning business during my middle and high school years. They also had one commercial client in the insurance industry, which is their most memorable client for me. It was the same local life insurance carrier agency office Mr. Larry Eubanks worked for. I had spent much time in the office growing up, helping my grandparents clean it up after hours. I had always imagined what the office looked like every day with agents working in it since I had only seen it after hours. I imagined people on their computers doing big business and thought it would be cool to do that one day. I am amazed at how the world has led me to insurance all these years. Our family insurance agent was a vital part of our family. Even 30 years later, I still haven't forgotten his impact.

Mr Larry Eubanks is no longer with us, but his impact on me and many others lives on. His impact on my career has always been the driving force. I have always wanted to leave an impact on my clients during every interaction as an agent, broker, and consultant;

you never know who you could be inspiring. Many years would pass before I realized the depth of the sweet insurance man's impact on my life's destiny. So many people have great stories of an insurance professional who impacted their lives, so I'd love to read them here. It helps to remind them of the inspiration behind it all.

Tell me about an insurance professional who has impacted your life and work.

....

....

....

....

For the next two decades, I pursued what I thought was my destiny and dream job, and before you ask, no, it wasn't the insurance industry. Since the tender age of 4, everyone around me knew that they knew what I would do with my life; it was my first love, singing. I have been a professional studio vocalist, performer, and songwriter since I was 14. People knew me in the South for being the young girl from Dothan, AL, who was on a little TV singing competition show, "American Idol." I was 15 when I went on the show, Season 2, and I made it to the Top 32 in Hollywood before the live shows. My hometown would tell you " Cierra Holt" {my former name} will be a famous singer. I was voted "Most Likely to be Famous" during my senior year in high school by all my classmates, so imagine everyone is surprised that I ended up in insurance and insurtech of all industries. I chuckle on the inside every time I think about it.

As you can imagine, I hated being a struggling artist. I had already grown up poor, so I wasn't interested in continuing my struggle in music. I pursued music for 15 years before I stopped to focus on something more significant: the birth of my first child. When I looked into my sweet boy's eyes at age 23, I knew I didn't want him to ever understand what struggle was, so I had to do something, and I had to do something, and fast.

After he was born, I couldn't find a job for a whole year while living in Atlanta; he was born during the 2009 recession, the worst time ever to be unemployed with a newborn. Working in retail, two jobs at a time in a lingerie and sneaker store in Dothan, AL, before my son's birth, made me know I had a knack for sales and talking to strangers. So, I returned home to my small town of Dothan, AL.; I was a failure, a single mom with no job and no degree since I left school to help take care of a sick family member before all of this.

I started working in retail, but when I returned to Alabama, my family poured so much hope and inspiration into me that I had no choice but to figure out my life and be successful no matter the obstacles I faced. My Grandma and Grandaddy allowed my son and I to live with them for free so I could focus on what I wanted to do to provide a good life for us. I returned to retail with a different mentality during that time. I started working at a wireless cellular store, and to my surprise, I was killing it; I had turned a $ 10-an-hour job into a $50,000 in 6 months because I made so much commission. I started exploring moving back to Atlanta and transferring into a corporate sales role.

I had taught myself everything at this retail store. I taught myself marketing: I printed off the coupons for the local fast-food restaurants, businesses, and corporations that gave out employee discounts on cellular service and would pass them out with my card in the drive-thru and everywhere I went. I would print and pass them out, teaching myself door-to-door sales. I also had a vast referral client base; someone was always waiting for me every hour to work with me and me only.

One day, things shifted for me; an insurance agent worked next door to my cellular retail job. She came into the store one day to visit with me, as she did often, but I could tell today was a bit different. She was a little weary because her agency wasn't going well and she was looking to hire more support staff to help her build a successful agency. She stated, "Alexis, you have sold a phone to almost every person who has visited my office; you are a natural relationship builder. She asked, " Have you ever considered becoming an insurance agent?".

Before I had my son, I had considered the insurance industry. I had even taken an assessment test for two insurance carriers and scored off the charts, but at that time, you could only work in the insurance industry for big carriers if you had good or decent credit. Unfortunately, I lost everything during the recession and had to start over completely. This time seemed meant to be for me because I had just started rebuilding during that last year when she asked. I finally had good credit again, which allowed me to take a chance on an insurance career again.

So, I met with the recruiter she introduced me to from corporate and went through the whole recruiting process. I took the initial assessment, and I scored off the charts again. It was not surprising, but then I passed the credit check. I started building a business and developing a marketing plan to pitch myself to the leadership team for this significant career as a captive agent. I had no idea how to do any of this stuff; I taught myself everything and leveraged the soft skills I developed by being in beauty pageants and talent shows on how to capture an audience and sell a solution but in a consultative way.

I also asked my Uncle Roger, the only other member of my family who has been an insurance agent for over 50 years, for advice. My Uncle Roger was so proud that someone wanted to pursue the same career that fulfilled him all these years. I also asked successful agents in my community for advice and insight on what an insurance career would look like for the first 2-5 years as I built a scratch agency in my hometown. It became clear to everyone that under the agency model, I was pursuing—a storefront-style agency with a staff of at least four people—I wouldn't be able to pay myself for at least the first two years. I had no idea how I could manage that as a single mom, and that's when I realized that most agency owners under this model were married or single with no kids, so they had extra support. I then asked my clients at the cellular store if they would support me if I started my agency. Many of them said yes, but many also said, your competition is the only other black woman agent in the state for this career, and most of my natural market was already her clients, which I wasn't prepared for. They all recommended that I get insight from her before making my final decision and presenting to the executive team about this career.

So I called Thelma. I told her my plans and asked if I could meet with her to discuss how to approach a career in this industry. I was also studying for my property and casualty license at the time. Before making such a huge transition, I decided to invest in myself by researching and taking the first license. I wanted to ensure it appealed to me, get a feel for what I would sell, and grasp the information well.

I prepared for this meeting for an entire week. I spoke to a few of her clients and community members who knew her to understand her personality and work style. I also talked to my Grandma and Granddaddy about her because they had been her clients for over 10 years. She was the first African American woman in Alabama who had an agency with this major carrier, so I wanted to come prepared for this meeting. I also wanted to ask her to mentor me because she was at a level that I aspired to be at in my insurance career.

I pulled up to her office in my beat-up 1989 Honda Accord with the front bumper barely hanging on with a wire hanger that my Uncle Reco rigged for me so that it would stay on the car. I was embarrassed about my car, so I generally parked far away, but I parked closer to the door since there were only two front spaces left in the lot in front of her agency. I went inside and spoke to the sweetest ladies in her office, and they let me know that she should be arriving any minute now. I heard a sports car engine revving and looked out the window.

There she was, gliding out in her two-door sports car, wearing a fancy business suit and sporting her shiny wedding ring. I thought, "That's her, looking like money." Then I thought, "I want to look like money, too, when I'm an agent." This excitement about my future as an insurance agent only increased.

She called me into her office, which was very fancy; she had many awards on the wall, such as Legion of Honor, Bronze Tablet, Silver Scroll, and Golden Triangle awards, along with articles of good work she does in the community. I had done my research, so I had a ton of questions. She asked me a few questions, and then I started with my list. She stopped me midway through maybe my third question and said, "Alexis, have you even passed the P&C exam yet? My girls have tried multiple times and never passed, so maybe you should focus on that and less on all these questions you have about your career as an agent". I was taken aback by that but understood her reasoning. I said, "Ms. Thelma, I will be taking the exam next week, and I truly believe I'm ready for it." she said, "Okay, well, let's see how you do. Then you can ask me more questions." That seemed fair, don't put the horse before the cart type thing, so I agreed and said, "I will call you as soon as I take the test to let you know how I did and then we can talk again.."

One mistake I made was telling her about all the resources this particular carrier offered new agents. Since she didn't get that type of support up front, she felt like I wouldn't work hard enough like she did to build a successful agency since they were giving so many incentives for success. I understood her point and made a mental note, but I was a little discouraged about my upcoming P&C exam since she said her CSRs took the test multiple times and didn't pass. However, I was determined to provide a better life for my son and me.

Despite the negative feedback I got about the new career I was pursuing, I studied from sun up to sun down. There was a lot of negativity, but thankfully, my family was so proud of me that they kept pouring positive affirmations and prayers into my spirit to drown the negativity coming my way. My Grandaddy would often say to me, "Cierra, you've been at the bottom long enough; it's time for you to go to the top. I've seen it in my dreams multiple times; people are going to know your name all over the world". My grandma and grandaddy have always had a way with words; I believe it's a pure gift.

I took the P&C test in Montgomery, AL. I will never forget how I felt the day that I took that test. I got to the last question and clicked the button to grade the test. My heart beat so fast, and the screen read, "Congratulations, you have passed the Alabama Property and Casualty Exam." I wanted to scream and cry and cry and scream some more. I was so excited, I did it. I immediately said I wanted to do it again, so as soon as I got home, I paid my money to that same online school for the Life and Health exam.

I remembered I needed to call Ms. Thelma, so I stopped what I was doing and called her agency to let her know the good news. She was so shocked that I passed the exam and was even more amazed when I said I had already registered for the Life and Health exam so I could knock that out too. I didn't have a contract or appointment with any carrier or agency; I just had much faith in myself and my ability to be great at insurance. She asked me to come to her office the next day so that she could talk to me more about my new career as an insurance agent. This was a different tune than she had before, but I was happy to oblige because I still needed a mentor.

I went to Ms. Thelma's office the next day. I was expecting to ask her about my extensive list, but she stopped me to make an offer. She offered me her agency when she retires in 5 years, which is the key to the kingdom in my eyes. I would run the agency in my small hometown of Dothan, AL. I felt I shouldn't answer immediately, so I didn't. Think it over and continue your process with the carrier directly so that you can evaluate both offers at once. I told her I would think about it and continue my process. I had already spent money and time away from my toddler building the business and marketing plan, so I needed to see it through.

I went to Birmingham, AL, for the agency candidate dinner and a presentation to the executive team. I will never forget the energy in that room at my first dinner as a licensed insurance agent. The candidate pool had many people from various backgrounds considering transitioning into an insurance career, just like me. There was a former MLB player, a social worker, a realtor, and even a lawyer.

There I was in the mix, a former singer and retail employee finally taking a career in the insurance industry seriously.

I presented the following day to the board, and they were so impressed that I made it to the final stages. Then, they paired me with a local agent to work with so that I could get on-the-job and field training, which I appreciated. At this point in my educational journey, I knew that I learned best in a hands-on training environment. However, there was a catch! I also needed my life and health license to work in the agency for a week for the best training experience. So I scheduled my license exam on faith on Friday before my start date and studied Monday-Thursday. As many of you know, you can't sit for the exam unless you can pass the practice exam in your pre-licensing course, so I studied 18 hours a day and had my family help me with my toddler at the same time until it was time to go to take the exam on Friday. I passed the practice exam at exactly 11:55 pm the night before the exam, took the test the next day, and passed on the first try. I was elated because I could train with an agent on Monday.

I got to work with another great individual who greatly impacted my career: an extremely successful local realtor in my hometown, Lan Darty. Lan is one of my top three favorite and most impactful bosses. He led with empathy, passion, and knowledge. He always makes his staff feel appreciated. He would start every staff meeting on Monday morning, allowing everyone to talk about what they learned in church or their place of worship over the weekend because he welcomed diverse beliefs and backgrounds.

I'll never forget one day the schools closed down mid-day for one of those "it's supposed to be bad weather days, but it's sunny outside type days off from school." He told us not to worry about it, to bring our kids here, and to set up the kitchen with a TV and games. I often told myself this is the type of agency owner and leader I want to be, so my staff always knows I value them. I did pretty well, too; I sold many policies in property & casualty and life & health. I was learning so much about insurance through this hands-on approach.

I vividly recall one day when a client called to cancel the insurance on her elderly mother's car because she no longer drives. One of the other CSRs in the office asked the client if she'd consider selling the car instead of leaving it undriven in the garage. The lady said it has sentimental value, so it would have to be for the right person who would appreciate it.

The next day, the client returned to the agency, and they asked me to come outside to see it.

They told the client about my story, what I was driving now, what I aspired to be, and how I should drive something that matches that. The woman sold me the barely driven/like new 1998 Toyota Corolla with only 50,000 miles on it for $1,000. I couldn't believe it! I cried tears of joy when I sat in that car with the a/c, a radio that worked, power windows, and a bumper that looked great. I was so grateful for that moment.

Lan had become the mentor that I was looking for. He had evaluated the market we lived in just as I did, and there wasn't a lot of growth potential. My competition would have been the agents who all worked for the same carrier, not the other independent agents or carriers in my hometown. He understood my income situation and told me I should explore other agency career models. He added one that paid a base salary with commission, paid for the office, and didn't require such a large staff or overhead because he saw how successful I could be if I had the right agency model. He also advised that I try other models before settling on which I did.

It was the best decision I could have made because I chose a model that allowed me to relocate back to Atlanta and start an agency career with no upfront cost; it even allowed me to attack one more set of licenses. I became not only an insurance agent but a financial representative when I acquired my Securities Series 6&63 Licenses, which was by far the most challenging test I have ever taken in my entire insurance career. I passed them even after a stressful retake because I didn't sleep enough and drank too much coffee the night before the exam. However, I don't recommend that before any insurance or financial exam.

While studying for my exam, I tried being a life insurance-only agent, and that's when I fell in love with accelerated death benefit riders added onto life insurance. When I passed my securities exams, I got deep into all aspects of my insurance agency career and landed the perfect model for my full-service insurance career. I knew then that I couldn't work for Ms. Thelma; I needed to start my agency. I was the first single mom that they had ever hired under this model, but I came highly recommended by an agency manager in Alabama that I interviewed with. I was sincere and said my goal was to get back to Atlanta, so he referred me to an Atlanta agency manager.

I remember my final interview. The carrier was worried that Atlanta's agents and financial representatives had large natural markets of at least 2,500 people. Still, I convinced them that the 25 people I knew would help me write higher quality business with nice premiums to match. I sounded like a seasoned insurance agent because I had interviewed with or had informational sessions with agents at every level of their career. Hence, I knew exactly how to hook them, but I followed it up with a strong book of business that I would be building from scratch. I had a little book of business being spread across all the agents in my office, so it wasn't from scratch but close enough. Before I knew it, I had grown that book of business in Alpharetta, Ga, from 25 clients to 1,165 client households within less than 2 years, all on building relationships in the community. My agency got 75% of referrals for new business and had 11 car dealerships that sent me business monthly. I worked until the dealership closed, some nights at 10 pm, when the other agents were not answering their phones because it was after business hours for them. I was a hungry single mom, so I was willing to do the work no matter the time of day. I was bringing in a lot of new business for my agency via auto insurance policies and then cross-selling them the next day to other products and solutions, which helped me quickly grow my business book. I had gotten so good that the carrier I was captive with at the time even had me train other agencies at the corporate level on how to do dealership partnerships effectively. The carrier even implemented my idea company-wide as a great way to bring in new business via partnerships with car dealers.

Then, one day, it all paid off. By the end of year 1 of my agency, I was named by an elite Atlanta and Chicago publication, known as "Rolling Out" magazine, as one of the "Top 25 Most Influential Women in Atlanta" with some pretty big movers, shakers, and legends. People knew me because I focused on educating my clients rather than selling them policies, and because of that, I became a 75% referral-based agency. People believed in my work and vision for insurance education in my community, so they supported me. I never purchased a single lead during my 4-year tenure as a captive agent with this carrier. Rolling Out magazine wrote an article on me after winning the award, and a high school in Chicago created an essay contest inspired by my story of " How to create a 5-year plan for success like Alexis Holt". I was so honored; I never thought that I would make it in the business for 5 years, so that's also why I never wanted to accept Ms. Thelma's offer because 5 years seemed like forever to me.

Still, I'm so glad I said no because look at what I accomplished in three years of starting my insurance career. I would've been holding myself back from success, waiting to benefit from someone else's success and hard work. I had never even been to Chicago when this happened; the story just got so much traction that it gave me an excellent opportunity to give back to them for honoring me. The carrier I was with matched my donation, and we awarded the winning essay contest student a scholarship.

A year later, I purchased my first home as a single mom, thanks to all the success of my agency, and moved my office to Dunwoody around this time as well. People saw me as a trusted advisor in Alabama, Atlanta, Chicago, and many other places. At the time, I had a theme for my career. I would end every social post with #Forbesbound because I felt that this was the beginning of my insurance career, and it could only get bigger and better as time passed.

After I purchased my first home, a five-bedroom, four-bathroom, new-construction home on a lake in metro Atlanta, I had many single moms reach out to me for advice on transitioning to a career in insurance. I knew that there weren't a lot of companies that would know how to how to work with a lot of single moms, so I decided to go independent. I started my own MGA (Managing General Agency) with 26 single moms to whom I had the privilege of teaching the business. The women were in Alabama and Georgia, and I had an overflow of interest, but I agreed that I wouldn't go past 26 women since I was 26 years old when I finally said yes to an insurance career. Hence, it held significance for me in so many ways.

That was also the year I officially became a single mom with a toddler who trusted that his mommy would provide for him, and I wanted to show other women how to do the same thing. I wanted to show them how they could build wealth for their families with an insurance career and be the trusted advisor in their communities. We did so well- we had women who used to be flight attendants, physical trainers, sales managers, social workers, and more. They all knew how to build relationships, which was key to their success, and I was happy to teach them everything I knew to help them be successful—the agency we built from scratch for almost 6 years.

I always planned for them to eventually branch off independently because I wanted more for them than just working for me. I tried to impact the industry outside my agency walls significantly, so I needed to sell and do something different. After selling my agency, I wasn't sure what was next for me, but I wanted to take time and figure it out. Then, I met the man who would help me get to my next level right after I sold my agency, Marvin Vaughn.

I met Marvin when I was trying to redefine who I was career-wise after agency life; I was exploring all types of roles at the corporate level so that I could still interact with agents because helping them was still my passion. I know you're asking yourself, what about Marvin, that helped you reach your next level? What did he have that was so special? He was a Talent Acquisition Manager, and he understood talent and job profiles like no one I had ever met in my entire work life. I asked him for advice on what I should do next with my career. I told him I now know a lot of people thanks to the agency reputation I built, so how do I get paid for having a vast network as a career? He said that's business development; they are relationship builders who also sell but focus more on managing the client relationship.

He also told me you should explore corporate insurance roles. To be #Forbesbound, you must have sold insurance on a corporate and global scale. He then explained the difference between an insurance agent and an insurance broker. My consultative approach to sales would transition well to a broker doing business with midsize to large organizations globally. He then told me the importance of my LinkedIn profile; as an agent, I used LinkedIn, but less than other social platforms, because I was selling directly to consumers. He told me that if I desired to attract big brokerage firms, my LinkedIn presence needed to improve and that my profile should match my resume while exploring new opportunities.

I made all the changes he suggested, and within 2 weeks, I got a call from a headhunter telling me about a role for a Big Four firm; he was right. I knew then that I would marry this man because he gave me game-changing information within the first 2 months of dating him. This is why it's important to date or marry someone who will always lift you and support your ambitions and not see you as competition but only focus on elevating you while they elevate themselves. I was sold on him, but that's another story for another book.

I landed that job with the big four firms and became a Senior Associate with a 6 figure salary and 5 figure guaranteed bonus to match. I could not believe I landed that because of Marvin. He even talked me through the importance of negotiating because women sometimes feel uncomfortable doing that. Now, I negotiate everything: Salaries, Titles, Job Descriptions, and Clients. It is all negotiable, and I'm thankful he taught me that. I did well at that Big Four firm and even led the Women's ERG as the Chair for the Atlanta office, all in my first year.

I missed being in a smaller firm, where I could be more impactful with my work and make changes more flexibly when things weren't going well in a deal.

That's when I discovered the Insutech space. I landed a job made for me in so many ways that the job description even included my hobbies. I got to work the way I liked then, wearing multiple hats. I was hired by a cyber insurtech as the Director of Agency Marketing. I got to build my division from the ground up. I was working with both agents and brokers, two groups of insurance professionals I completely understood since I had been both of them in my career.

They only had 4,500 agents and brokers when I came on board, but I understood immediately because I put myself in their shoes. I said this is a brand new product, so we need to train them and the consumers on why they need to sell it and why the consumer SMEs need to buy it. I became a continuing education instructor and worked with another one of my top 3 bosses in my career, Isabelle Dumont. She took the #1 spot all day, every day! She was the most intelligent person in the room every single time and never flaunted it. It was inspiring to work side by side with her; she saw me as an asset and not a competition, so I taught her insurance. She taught me AI and Cybersecurity; the dream team is what I call us. Together, we helped grow that MGA from 4,500 agents and brokers to 23,000 agents and brokers in less than 2 years with a significant focus on educating the agents, brokers, and SMEs.

We did strategic and creative marketing campaigns that always moved the needle tremendously. We created partnerships that saw immediate results within the first few months. We put together events that were curated experiences tailored to each audience, and I got the fantastic opportunity to be myself in every aspect of my career. I also hosted and produced a Top 20 Apple Business News podcast where I interviewed the FBI, Former CIA agents, Insurance Agents, Brokers, and Cybersecurity experts. I got promoted 3 times in two years, all the way to AVP. She helped me build my executive and speaking profile by encouraging and allowing me to live out my dreams.

Because of Marvin Vaughn and Isabelle Dumont, I felt the courage to start my own company after leaving the cyber insurtech. It has turned out to be wildly successful in a short time, just like my agency career, because I added a mix of what they taught me and what I have learned in my career, took my resume, and built a successful company. I added some hot sauce by also bringing my now husband Marvin Vaughn on as a Co-Founder & COO of Off Course, leveraging his background in talent and consulting at the insurance carrier level, aviation, and healthcare-

We, indeed, are the perfect pair in life and business. We know our strengths and operate in our business that way. Together, we run a successful global consulting firm with multiple divisions, Off Course Consulting, Learning, Media, and Tech, in Atlanta, Ga, and the United Kingdom, in Cardiff, Wales. We now work with insurtechs, MGAs, carriers, reinsurers, insurance agencies, and brokerages, getting them back on course with Off Course through strategy consulting.

I would've never imagined that 15 years later, I would still love insurance as much as I did music. I was born to do this work; it's bigger than me. The impact has been tremendous thus far, and I can't wait to bring more people into this industry that I love so much through the great work we are doing at Off Course. Now that we are continuing my insurance story, I hope this helps set the stage for you on why I have a unique perspective and pulse on the future of insurance distribution for agents and brokers.

REFLECTIONS

Sometimes, it helps to revisit your insurance story, especially during tough times when business isn't going as planned. Let it serve as a reminder of why you chose to enter the insurance industry in the first place. What was your original motivation? We know that this can change over time. Share your story with us...

What was your original vision for your insurance career?

....

....

....

....

What are you doing now?

....

....

....

....

Where do you want to go next in your insurance story?

....

....

....

....

CHAPTER TWO

SPEAK THE LANGUAGE: THE ABCS: AGENTS, BROKERS, AND COUNSELORS

This chapter is for people new to entering the insurance industry and agents and brokers trying to figure out where they want to go for the next level in their careers[1]. I'm often asked by insurtech and fintech founders new to the insurance industry what the difference is between an insurance agent and an insurance broker. That's when I realized there are also a lot of agents and brokers who don't know the difference, which is expected since, typically, agents and brokers stay in one lane for most of their careers, so I'll break it down for you. I'll give you actual definitions along with my understanding of what they do similarly and differently.

My UConn students do a great job of explaining the difference now that I've also taught them. One of the things I start with is understanding that not all audiences are created equal. The values in a solution, product, or service may be different for both agents and brokers. I will explain how they differ. I'll start by saying this: agents are visible in the community, while brokers are often in the shadows, working with businesses and high-network individuals.

Distribution Breakdown:

- **Insurance Agent**—An Agent works with and for the carrier. They are representatives or employees of an insurance provider.

An agent must be licensed to sell the products their companies promote. Insurance agents connect the policyholder and the carrier and are generally held to a "suitability" standard. This means they must ensure that the client can afford the insurance offered.

- **Captive Agent**- An agent that works exclusively for one insurance provider; this could be with an agency or carrier.
  - By choosing a recognizable insurance provider, captive agents can access training, brand recognition, company benefits, and other perks.
  - Some carriers and agencies even offer a salary or draw, which is helpful when building a new book of business or cross-selling one you may have purchased.
- Captive agents are exclusively appointed to sell the insurance products with their carrier or agency, with which they have an exclusive agreement.
  - Some insurance providers will allow you to own your book of business. However, they will only allow you to sell it to a current or incoming agent who is also a captive agent but never to an independent agent outside the organization. Also, remember that some captive carriers and agencies will not let you own your book of business; you have to leave it when you leave the company, and they will distribute it evenly to current/in-house agents, and you get no say in that.
- **Independent Agent**- An agent appointed by multiple insurance providers to offer more than one insurance product to their clients.
  - Independent agents don't receive training unless they have decided to work with an insurance provider that offers explicit product training on their products. Independent agents usually have to train themselves, so I don't advise this model if you are brand new to insurance. Learn the basics first as a captive agent, and then

consider going independent later only if it makes sense. They also don't receive benefits and must build their brand as an agency and small business. We will talk a bit more about that later.

- **Insurance Brokers**- Brokers work with and for the client for compensation in broker fees, commission, or a combination of both. They shop with multiple insurance providers to find the right solution for their clients. They often get paid by the client, and depending on the insurer's agreement, they can also get paid directly by the insurer. Most brokers are licensed and have binding authority, but the carrier or wholesaler typically binds the coverage depending on the circumstances. They are impartial regarding products and do not have the authority to "bind" coverage on an insurer's behalf in every situation. Still, they do have binding authority in some carrier agreements. Insurance brokers must operate in a fiduciary capacity, a higher standard of responsibility that obliges them to prioritize their client's best interests. A broker is more of an insurance consultant who works on behalf of the client. A broker must still be licensed to negotiate insurance products with which the company works.

  - This role often requires an additional license as an insurance counselor in certain states. It involves a deeper dive into certain products so the broker can advise the client on the policy details, which generally isn't an agent's role. They must also be licensed to do the job but won't be appointed directly to sell the solution like an agent would.

  - Brokers often make a salary plus commission and receive benefits as employees of an insurance brokerage firm. This varies depending on how long they have been with the company, the area of expertise they consult in, average deal sizes, client sizes and types, and various other factors.

- **Insurance Producer-** a producer has combined functions of both an insurance agent and an insurance broker, according to the NAIC's definition. No matter what, the insurance producer must be licensed in that (LOA) line of authority they are selling, soliciting, or negotiating. In other words, insurance producers are generally considered to be acting as insurance agents—and are typically subject to a suitability standard—if they have an appointment with an insurance carrier. However, if they are acting in a capacity that does not require an appointment, they are held to the higher fiduciary standard as insurance brokers. This has only been adopted in certain states, not all.

Thinking about the nuances of the insurance distribution side of the industry, I have had the privilege of having a mountain-top view of the distribution business. I decided that I would explore every area of the insurance industry that my career would take me instead of taking the traditional route of staying in the exact location for my entire insurance career. When I got to the top of the mountaintop, I could look down and see the world of insurance from a different viewpoint; I could see every single part of the industry. Due to my distinct view of the industry, I've been exposed to various structures and compensation plans on both the agent and broker sides of the industry. I felt that it was essential to break this down for you because if you're new to the industry, you would want to know this information without having to sit in hundreds of interviews to figure out where to start with an insurance career on the distribution side. Let's start with the insurance agent. I'd like to break down the captive and independent agents for you.

Here is a breakdown of what I learned about the variations of each insurance avenue:

Captive agents have a few different options, so let's start with what we typically see in the industry and go from there. A captive agent can be a W2 agent or a 1099 Contract agent exclusive to a carrier or an agency.

Captive Agents can typically have the following nuances:

- **Scratch agent**-build a new business client base from 0-1,000 via purchasing leads, building community relationships, etc.

- **Acquired agency or book of business**-purchase, a book of clients from an agent or agency exiting the insurance industry through either retirement or career transition. You can cross-sell the clients within that book to new businesses and make an income from the renewals of that book of business.

1099 Captive Agency Owner Model:

- Some captive models, the agency owner is a 1099 contractor that holds the appointment with the carrier exclusively.

  - This model can also require the agency owner to hire a team that pays a W2 paycheck or 1099 hourly rate or salary payment. These employees must be licensed to sell the line of insurance your agency sells or solicits to interact with clients. Still, each employee can also have one license and specialty focus within the agency. For example, an agency owner can have one employee licensed to sell only life and health insurance, and that person will only solicit or sell life and health products. However, the agency owner must be licensed in all lines of business that they are appointed to deal with the carrier since they typically hold the carrier appointment in their license name on behalf of their agency.

  - This model also requires a significant investment in the business from the agency owner:

    - This could be in the form of a minimum investment or access to capital required by the carrier, such as $100,000, which allows them to gain equity in their agency or acquire a book of business from a retiring agent in some cases.

    - There could be a requirement to pay the ongoing cost of leasing a storefront location for the agency. In some cases, carriers will also offer a startup stipend via a line of credit with their bank or cash, doubling as the carriers' investment in your agency's success and covering their equity [AH3].

- Paying salaries or hourly rates for a team is a significant part of your investment in this model.

- You could also be required to relocate your family based on the availability of the agency territories that have openings when you come on board. Carriers typically prefer you to remain within your same state or city since most of your natural market may also be there, which helps when trying to obtain new business.

- In most models like this, you typically hear from agency owners that they cannot afford to pay themselves the first 2 years due to the high overhead cost they incur, so having a spouse who can afford to carry you during the first two years while you build is critical. I would not recommend this model to a single person or parent if you don't have access to significant capital that you can put your hands on quickly.

- Within this model, an agency owner typically sticks to a specific territory regardless of whether they are licensed in the whole state. For example, suppose your agency is located in Alpharetta, GA. In that case, you are typically only allowed to write business within the allowed zip codes and county lines so that you don't step your other colleague's toes by selling in Newnan, GA, over an hour from Alpharetta. This also helps with getting clients who live nearby to drop into the office for meetings and change requests for their policies. Sometimes, in this model, your biggest competitor is your colleague because you can't take a family member who is already their client to be your client just because you opened an agency. An agent's tenure can also be a factor here; I know some agents who were allowed to write business all over the state because they have proven that they can write quality business, but this wouldn't work for a brand-new agent.

- Some carriers offer a brokerage service run by either another department or a sister company of the airline. Still, you are typically only allowed to write with the brokerage department as a captive agent if your client gets declined first by the carrier you represent. Then afterward, they could use brokerage for reinsurance or find a nonstandard carrier similar to an independent agent so that they don't lose the business altogether.

- Most captive models still require the agent to be licensed in all lines of business that the carrier or agency represents, including securities licensing. I have also seen models where they let the agent choose to be purely an insurance agent or a one-stop shop financial representative who is also an insurance agent, which means you may have both insurance and securities licenses.

- Credit can also play a significant factor in this model, especially since decent credit is required for a securities license, and you need credit to get commercial leases and acquire supplies and furniture for your storefront agencies.

- You will receive training in this model; this model often recruits agents who are brand new to the industry or may be looking for a change. Carriers and agencies typically offer pretty comprehensive training, which is usually paid. Even if you travel for the training, they typically have no problem making a training investment in the agent to ensure their success. For example, the training could be sales-related or product-specific, and they may even pay for the agent to get licensed if they aren't already.

- In this model, you may only receive benefits during the training phase, and then you and your staff will no longer receive benefits or salary assistance once training commences. It will be up to the agent to decide if they would like to offer benefits to their staff after training; also, it's important to note that some models don't offer benefits to the staff or agent at all. The types of benefits also vary by carrier and agency.

- Agency owners are typically 100% commission regardless of whether they are a scratch agency or an acquired agency.

Let's dive into the W2 Captive Agent Model:

W2 Captive Agent Model:

- In the W2 Captive Agent model, the agency owner or agent is an employee of the carrier; they normally receive perks like benefits and training.

- In this model, a salary or draw is offered to the agent for a period of time anywhere from 6 months to 3 years. You are typically still allowed to receive commissions on top of this guaranteed amount until you transition to the 1099 agency owner model, which will be 100% commission.

- In this model, agents are not required to hire a team. Still, they can hire someone on either a W2 or 1099 appointment and even split the cost with the carrier or other agents with whom they may share an agency office to access a service rep or sub-producer to help with servicing or handle additional sales to the clients.

- Agents oftentimes share an agency office with other agents in this model for carriers, or a captive agency may not have an office at all, so a home office could also work since you are the sole salesperson and service person typically in this model.

- For most employee agents, the overhead cost for having an agency can typically be taken care of by the carrier or agency, so you wouldn't need to make a significant investment in a storefront until you reach the portion of their agreement where you transition into a 1099 contractor agency owner. I have seen the transition periods be anywhere from 6 months to years, depending on how that particular carrier structures its model.

  - In this model, the agent appointment is set up exclusively with either the carrier that the agency has appointments with or exclusive to that carrier's in-house products only, and the agent cannot sell outside of those products.

- In this model, the agent can request a book of business at no cost to them if it's available, or the agent could build an agency from scratch as well.

- This model also requires specific territories so that you don't impede on your colleagues who also service nearby locations. This could be by zip code, county line, etc, regardless of being licensed for the whole state.

Insurance Agent Compensation:

- **Captive Agent 1099 first-year compensation**

For a 1099 contract insurance agent, compensation can vary widely depending on commission structure, overhead costs, staff expenses, and operational requirements. 1099 captive agents often bear the cost of their own office space, staffing, and other operational expenses, which can impact net income. Compensation ranges are gross income before expenses. Net income will depend on total overhead costs.

Here's a breakdown of potential compensation and expenses across various career stages: Here is a detailed breakdown of compensation for 1099 captive insurance agents, including Property & Casualty (P&C), Life, and Health insurance. Compensation is segmented into new policies, renewals, and annual earnings by years of experience.

Property & Casualty (P&C)

New Policy Commissions	5% to 10% of the premium
Renewals	2% to 5%, up to 15% with select carriers
Annual Compensation Progression	First-Year : $40,000 - $60,000 5 Years : $60,000 - $80,000 10 Years : $80,000 - $100,000 20 Years : $100,000 - $120,000

Life Insurance

First-Year Commissions	40% to 120% of the first-year premium
Renewals	1% to 2%, with some carriers limiting renewal commissions after a few years

Annual Compensation Progression (including commissions)

First-Year	⟶	$50,000 - $70,000
5 years	⟶	$70,000 - $90,000
10 years	⟶	$90,000 - $110,000+
20 years	⟶	$110,000 - $130,000+

Health Insurance

First-Year Commissions	20% to 30% of the first-year premium
Renewals	3% to 10% for long-term policies, though many carriers limit renewals to the first 2–3 years

Annual Compensation Progression

First-Year	⟶	$30,000 - $50,000
5 years	⟶	$50,000 - $70,000
10 years	⟶	$70,000 - $90,000
20 years	⟶	$90,000 - $110,000

Combined Compensation (P&C + Life + Health)

Annual Compensation Progression	
First -Year	$120,000 - $180,000
5 Years	$180,000 - $240,000
10 Years	$240,000 - $300,000
20 Years	$300,000 - $360,000

1. Health Insurance Distinction: Health insurance generally has lower first-year commission rates than life insurance but can provide significant renewal income, especially for long-term policies like Medicare Supplements or Affordable Care Act (ACA) plans.

2. Combined Earnings Assumptions:Combined compensation assumes agents sell an equal mix of P&C, life, and health products.

3. Renewal Impact: Over time, renewal commissions for health and life products stabilize overall income, while P&C continues to offer consistent long-term renewals.

4. Market and Carrier Variations: Actual compensation depends on geographic markets, insurance carrier agreements, and individual sales volume.

Overhead Costs for a 1099 Contract Insurance Agent with Staff and Storefront

Staffing Costs (2-4 licensed W-2 producers working full-time)	
Hourly Rate	$18 - $30 per hour
Annual Salary per Producer	$37,440 - $62,400
Total Staff Cost (2-4 Producers)	$75,000 - $250,000 per year, Including benefits (e.g., health insurance, retirement contributions, payroll taxes)

Storefront Location Costs	
Commercial Rent $\longrightarrow$	$1,500 - $5,000 Per month (depends on location, size, and market)
Annual Rent $\longrightarrow$	$18,000 - $60,000
Utilities (Electric, Water, Internet) $\longrightarrow$	$200 - $500 per month $2,400 - $6,000 per year

Office Supplies and Equipment

Furniture (Desks, Chairs,etc.)	$5,000 - $10,000	(one-time setup cost)
Computers, Phones, Software	$2,000 - $5,000	(Per employee initially)
Maintenance / Upgrades	$1,000 - $2,000	(Annually)
Total Initial Setup	$15,000 - $30,000	(Annually)
Ongoing Supplies	$2,000 - $5,000	(Annually)

Marketing and Client Acquisition

Advertising (Online, Local Media, Events)	$500 - $2,000 per month
Annual Marketing Budget	$6,000 - $24,000

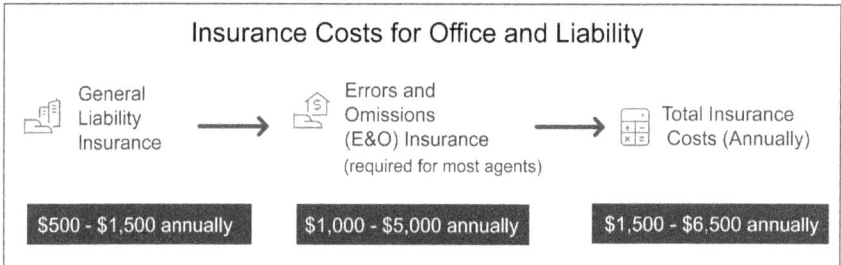

Insurance Costs for Office and Liability

General Liability Insurance → Errors and Omissions (E&O) Insurance (required for most agents) → Total Insurance Costs (Annually)

$500 - $1,500 annually | $1,000 - $5,000 annually | $1,500 - $6,500 annually

Licensing, Continuing Education, and Association Fees

Agent Licensing and Continuing Education	$500 - $1,000 per year
Professional Association Fees	$200 - $500 per year

Estimated Total Overhead Costs

Yearly Overhead (Including Staffing, Rent, Marketing, and Insurance)

Low-End Estimate	$100,000 - $150,000
High-End Estimate	$200,000 - $300,000

Example Net Income After Expenses by Career Stage

Year 1	Likely at break-even or low profitability due to high startup costs
Year 5	$40,000 - $80,000 in net income
Year 10+	$80,000 - $150,000 in net income
Year 20+	$150,000 - $300,000 in net income Depending on gross revenue growth and controlled overhead

These figures represent common ranges and would depend heavily on business volume, the efficiency of operations, and geographic area. A well-managed agency in a favorable market with strong sales could see higher profitability, while high overhead costs in a competitive or high-rent area could reduce net income.

- **Captive Agent W2 Agent compensation**

For W-2 insurance agents who receive a salary plus commission and bonuses, compensation can vary significantly depending on location, market size, insurance type, and individual performance. Below is a breakdown of potential compensation ranges for W-2 agents at different career stages.

Here's a detailed breakdown of compensation for **W-2 captive insurance agents**, segmented by lines of business. Compensation is divided into base salary, commission structures, and annual earnings by years of experience.

Property & Casualty (P&C)

Base Salary Typically $30,000 to $50,000, depending on the carrier and region
New Policy Commissions:5% to 10% of the premium
Renewals 2% to 5% of premiums, sometimes higher for select carriers

Annual Compensation Progression (including commissions)

First-Year	$50,000 - $70,000
5 years	$70,000 - $90,000
10 years	$90,000 - $110,000+
20 years	$110,000 - $130,000+

Life Insurance

Base Salary : $35,000 to $60,000 for life-focused agents
First-Year Commissions : 40% to 80% of the first-year premium
Renewals : Typically 1% to 2% of premium; renewal commissions often cease after 3–5 years

Annual Compensation Progression	
First-Year	$60,000 - $80,000
5 Years	$8,000 - $100,000
10 Years	$100,000 - $120,000+
20 Years	$120,000 - $140,000+

Health Insurance

Base Salary : $30,000 to $50,000
First-Year Commissions : 20% to 30% of the first-year premium
Renewals : 3% to 10%, often limited to the first 2–3 years

Annual Compensation Progression	
First-Year	$50,000 - $70,000
5 Years	$70,000 - $90,000
10 Years	$90,000 - $110,000+
20 Years	$110,000 - $130,000+

Combined Compensation (P&C + Life + Health)

Base Salary : Typically $40,000 to $70,000 for agent selling across multiple lines

Annual Compensation Progression (Base + Commission)	
First-Year	$110,000 - $150,000
5 Years	$150,000 - $190,000
10 Years	$190,000 - $230,000+
20 Years	$230,000 - $270,000+

Agents with Leadership experience

Salary : $60,000 - $100,000
Commission and Bonuses : $50,000 - $150,000+
Total Compensation : $110,000 - $250,000+
Senior Agency/Carrier Executive
Executive Salary Range : $175,000-$250,000+
Executive Bonuses : Varies; this could be a percentage of the salary anywhere between 20-30% typically, or it could be based on revenue generated by the organization

Total Compensation as an Agency Manager: $50,00-$300,000; this varies due to tiered commission structure based on the team goals such as gross written premium, chargebacks, qualifiers for presidents' clubs, and other awards, along with other various goals depending on the carrier or agency.

With ten or more years of experience, agents often reach senior status, and their compensation reflects a larger, well-established book of business. At this point, some agents have chosen the leadership route to train upcoming agents via agency management.

Commission and bonus earnings are likely substantial, especially if the agent works with high-value policies or has taken on leadership roles within the organization. Experienced agents are usually among the top earners if they've consistently grown their business and maintained client relationships.

In some cases, compensation can be even higher if the agent has advanced to a partner role or holds equity in the agency.

Independent Agent Model:

- In the independent agent model, you can choose one of the following:

- Have a solo agency where you do everything on your own, services work, and sales

- Be an agency owner that hires staff to help you service your clients or bring in new business.

- Build an agency model where you hire other agents as 1099 agents by giving them access to your appointments and taking a split of their commission.

- No startup money is needed to start an agency

- In this model, independent agents are compensated via a 100% commission

- In this model, independent agents can have appointments with multiple carriers, have a niche market focus, or offer multiple lines of business.

- The independent agent will need to seek companies who are open to appointing them as approved agents to sell their insurance products. This is often times accompanied by a quota in order to maintain the appointment.

- There is no territory restriction; agents can also sell products in multiple states with a non-resident license.

- An agent can have one agency license if they are only writing one line of business within their agency. For example, a P&C licensed agent can focus on one product, such as a commercial insurance agency.

- An agent has to either obtain their own direct carrier appointments or sign up with a wholesale brokerage that offers multiple carrier offerings.

- Independent agents can get quotes and bind policies with multiple carriers since they are allowed to have multiple insurance provider appointments.

Insurance Broker & Counselor Model:

- The insurance broker model is typically a W2 model, with no ownership in the book of business

- This model is paid in a set salary plus bonus or a decreasing salary plus commission, bonus, and, oftentimes, broker fees.

- Most brokers only focus on one insurance specialty. For example:

  - Property and casualty Licensed brokers and counselors may only focus on Commercial Products for businesses and High-Net worth and Private Client Lines, such as insuring high-value assets, like insurance for fine art collections, super yachts, collector cars, homes in other countries, and much more.

  - While a life and health licensed broker and counselor may only focus on group benefits.

- Brokers have an insurance counselor license that coordinates with either the specialty in property and casualty or life and health if they are in the group benefits space as well.

- Insurance brokers and counselors also have binding authority, but it's not a requirement for them to bind the policies when this can also be done by the carrier or wholesale broker, depending on the details of their contract, and it may even be deal-specific.

- They often have large compensation packages due to the size of deals and types of business they write.

- Brokers do RFPs and Proposals when going out for new business.

Independent Agent Compensation Structure:

The compensation structure includes base compensation, average commission percentages, and bonus ranges for Property and Casualty (P&C) and Life and Health (L&H) insurance agents. The breakdown reflects typical earnings at different career stages.

Year 1 (Entry-Level)

Property & Casualty (P&C)
Compensation Range : $30,000 to $60,000 (mostly commission-based)
Average Commission Percentage
New Policies : 8% to 15%
Renewals : 2% to 8%
Bonuses
Carrier bonuses for meeting sales targets : $1,000 to $5,000 annually
Volume-based bonuses : $2,000 to $8,000

Life & Health (L&H)
Compensation Range : $35,000 to $75,000 (commission-heavy)
Average Commission Percentage
Life Insurance : 60% to 100% of the first-year premium
Health Insurance : 20% to 25% of the premium (initial) and 5% to 10% on renewals
Bonuses
Sales-based bonuses : $2,000 to $10,000 annually

Year 5 (Mid-Level)

Property & Casualty (P&C)

Compensation Range : $75,000 to $125,000

Average Commission Percentage

New Policies : 10% to 15%

Renewals : 5% to 10%

Bonuses

Carrier bonuses : $5,000 to $20,000

Volume incentives : $5,000 to $15,000

Life & Health (L&H)

Compensation Range : $80,000 to $150,000

Average Commission Percentage

Life Insurance : 60% to 115% of the first-year premium; Residuals of 3% to 10% in subsequent years

Health Insurance : 15% to 25% initial, 5% to 10% on renewals

Bonuses

Carrier bonuses for persistency and volume : $10,000 to $30,000

Year 10 (Senior-Level)

Property & Casualty (P&C)
Compensation Range : $125,000 to $200,000
Average Commission Percentage
New Policies : 12% to 18%
Renewals : 8% to 12%
Bonuses
Carrier bonuses : $15,000 to $50,000 annually
Incentive trips and profit-sharing options : $10,000 to $40,000 in value

Life & Health (L&H)
Compensation Range : $150,000 to $250,000
Average Commission Percentage
Life Insurance : 70% to 115% of the first-year premium; Residuals of 5% to 15%
Health Insurance : 20% to 25% initial, 7% to 12% on renewals
Bonuses
High-volume bonuses: $20,000 to $75,000
Persistency and retention rewards: $15,000 to $50,000

Year 20+ (Veteran Agents)

Property & Casualty (P&C)
Compensation Range : $200,000 to $500,000

Average Commission Percentage
New Policies : 15% to 20%
Renewals : 10% to 15%

Bonuses
Carrier bonuses : $25,000 to $100,000 annually
Partnership/profit-sharing opportunities : $50,000 to $150,000

Life & Health (L&H)
Compensation Range : $250,000 to $1,000,000+

Average Commission Percentage
Life Insurance : 90% to 115% of the first-year premium; Residuals of 8% to 20%
Health Insurance : 20% to 30% initial, 10% to 15% on renewals

Bonuses
Carrier performance bonuses : $50,000 to $200,000 annually
Legacy bonuses and retirement benefits : $100,000+

To sum up the independent agent opportunity:

1. P&C Agents benefit from steady income due to policy renewals, with commissions from renewals growing significantly over time.

2. L&H Agents earn higher upfront commissions but rely on policy persistence for long-term residuals. Mid market, high net-worth

3. Bonuses and incentives grow as agents hit volume or retention milestones, with larger opportunities at senior levels or when partnered with carriers.

These figures reflect averages across independent agents and may vary depending on location, carrier agreements, and specialization.

Insurance Broker and Counselor Compensation Structure:

Here is a breakdown of salary, commissions, bonuses, and broker fees for insurance brokers and counselors working for the top 100 insurance brokerage consulting firms at various career milestones. The ranges reflect differences based on firm size, location, and specialization.

Year 1 (Entry-Level)
Base Salary : $44,000 to $89,371 Commissions
Range : 2% to 8% of premiums Example : For a $1,000,000 book of business, the commission could be $20,000 to $80,000 annually
Bonuses
Performance-based : $1,000 to $10,000 (dependent on reaching sales targets)
Broker Fees
$500 to $2,000 per transaction, where applicable and permissible by state law

Year 5 (Mid-Level)

Base Salary : $95,083 to $100,335

Commissions

Range : 5% to 20% of premiums

Example : For a $2,000,000 book of business, the commission could be $100,000 to $400,000 annually

Bonuses

Annual performance bonuses : $5,000 to $25,000

New client growth bonuses : $2,000 to $15,000

Broker Fees

$1,000 to $5,000 per advisory service or transaction, Depending on the complexity of the case

Year 10 (Senior-Level)

Base Salary : $104,379 to $125,000

Commissions

Range : 10% to 25% of premiums

Example : For a $3,000,000 book of business, the commission could be $300,000 to $750,000 annually

Bonuses

Annual performance bonuses : $10,000 to $50,000

Profit-sharing or partnership bonuses : $15,000 to $75,000

Broker Fees

$2,000 to $10,000 per transaction, with flexiblity based on expertise and service scope

Year 20+ (Veteran Brokers)
Base Salary : $120,000 to $141,000+
Commissions
Range : 15% to 30% of premiums
Example : For a $5,000,000 book of business, the commission could be $750,000 to $1,500,000 annually
Bonuses
Annual bonuses : $25,000 to $100,000
Profit-sharing : $50,000 to $200,000 (Depending on firm performance)
Broker Fees
$5,000 to $25,000 per transaction or advisory service

Residual Income: Brokers and counselors often earn ongoing commissions from policy renewals, providing a reliable income stream over time.

Specializations: Higher commissions are typical for life insurance and high-value commercial policies, while property and casualty policies may offer lower initial commissions but steady renewals.

Top Firms: Firms in the top 100 list often provide more structured bonus programs and higher earning potential than smaller agencies.

This compensation breakdown reflects an aggregated view based on industry averages and specific firm practices.

For agents who do not see this type of income range, either you aren't doing something right, or this just isn't your career path and that's okay. Maybe you should explore corporate insurance roles instead. We need you to stay in the insurance industry, so I would never direct you out of it.

Now that you have learned about the differences between the agent, broker, and counselor models, which model works best for you? And why?

Personally, I felt that transitioning from an independent agent to a corporate insurance broker was much easier since I was already working with multiple insurance providers, and those models from that perspective are similar, but where they differ is one can bind the new business, and the other can negotiate new business. Let's list the pros and cons of each based on your current and future lifestyle so that you can make an informed decision if you are new to the industry or transitioning to a different model.

✓ Pros	✗ Cons

CHAPTER THREE
UNDERSTANDING THE NEW DISTRIBUTION MODEL

Let's take a trip down memory lane to explore what insurance distribution has looked like over the years. We are all familiar with the agent and broker distribution model. I broke down the differences between agent and broker models in the previous chapter. The traditional insurance distribution model was more manual, linear, and reliant on intermediaries. Although it was under attack during the early years of the insurtech entrant, the agent and broker distribution model has been steadfast in explaining why the human touch is always needed in any insurance process, especially client acquisition, services, and renewals. For those new to the insurance industry, here is a look at where the distribution model has evolved.

The distribution model relied heavily on direct sales, which consist of two models: face-to-face and direct sales over the phone through call centers with licensed agents. The face-to-face model has been the staple of the insurance industry. This model includes in-person sales, field underwriting, and policy services being performed by the agent or broker. Agents typically meet with clients either in-office, at their homes, or businesses. This also depends on the agency or carrier model under which they are contracted. As a prior Captive W2 agent, I did all three since the carrier provided the corporate and agency offices. As a 1099 Captive Agent within the brick-and-mortar model, I only saw clients at the office. In both models, I handled the servicing of clients over the phone unless it was time for a renewal or (IFR) insurance and financial review time.

Still, typically, clients would come into the office to sign policy changes since insurers did not accept digital signatures from 3rd party document signing platforms for some time. As an independent agent, I did not have a brick-and-mortar office. However, due to specific licensing requirements, I was required to have a home office space and a way to store client files and paperwork. I spent all my time in the field with the clients, meeting face to face and only setting appointments over the phone. I also did drop-in meetings at businesses on the commercial and voluntary benefits side.

As a broker, I consistently reported to a corporate office that covered a specific territory and sometimes even particular products. We also often worked in silos. For example, you were either a commercial broker or a benefits broker, never both, because brokers focused on niche markets, products, and solutions. Very rarely will you see a broker doing everything unless it's their specific brokerage model, but when it comes to the top brokerages, often, they don't work that way. Brokers have different structures, but just like carriers, there is a base model from which everyone builds their brokerage.

I previously worked in a model where the broker had a team that supported their efforts in multiple ways. They may have an account executive who submits business on their behalf, a sales executive who hunts for new business, and the broker is usually the consultant seeing the deal through to the end, who is in all the meetings and presenting to the client. The broker could also have an RFP team that supports proposal writing for procurement efforts. I loved this part because they provided a template each broker could build on to ensure that their RFP always had the core information typically required from procurements with the appropriate language for the audience. For example, a government proposal will have a different language than a manufacturing proposal. RFPs are not for the weak; you must be able to stand firmly on what you put into those RFP responses because they can make or break the deal. Often, there are no opportunities for a do-over as you run multiple quotes and proposals or use the good, better, and best theory on the agent side of the business. You only get one shot at a deal.

Some insurance carriers used the direct sales model, selling policies directly to consumers through their in-house sales teams, but this was less common than the agent-broker model. Direct sales were limited to face-to-face interactions, call centers, or mail campaigns. Most carriers have a call center for sales teams, claims, and even underwriting in some models. Direct mail, TV Ads, print ads, and other traditional marketing campaign sales are also different models that are typically used to gain warm leads for agents in every vertical, which would prompt a client to either call into an office for a quote or set an appointment for an agent to meet with them in person, either in-home or in office, depending on the agency or carrier model.

Most information clients receive on a product has been limited to agents and brokers providing additional information. Customers could only fill out paper applications or in-house computer applications, which could only be done in person in the presence of an agent to gain insurance coverage. Agents typically did minimal physical inspections of a property by taking a picture on a digital camera to send it to underwriting for review. Insurance products were also one-size-fits-all and not individualized like most of our clients needed. Risk assessment was generalized, relying on traditional data sources like credit scores or broad demographic factors. The reliance on manual processes and physical infrastructure resulted in high costs, passed on to customers through premiums.

Since then, the insurance distribution model has evolved, encouraging agents and brokers to grow with it. In the new regime, agents and brokers are starting to take a modernized approach to servicing and selling to clients. People don't give agents and brokers enough credit regarding technology adoption, we are willing to evolve when needed. We have changed with the times because our customers are who we serve, and whatever we can do to make their lives better through the service that we provide as insurance professionals is what's important to us at the end of the day. It also has to make sense for our business simultaneously.

Digital Transformation for agents and brokers has happened in stages; they now use advanced CRM tools and digital platforms, and some even leverage AI tools to provide personalized recommendations and manage clients more efficiently. Omnichannel interaction used to be done only at the carrier level, but now it is done by agents and brokers through expanded communications that include virtual consultations, video calls, social media, and face-to-face interactions. I also want to note that people don't often consider that an agent or broker can only work in a way that is allowed by the carriers and regulators of these products. They control how insurance products are delivered, not the agent or broker. Agents and brokers can offer value-added services, allowing event agents to act as risk advisors and provide insights and services beyond just selling policies, such as financial planning or risk mitigation strategies often offered in the new insurance policies. I've also added a breakdown of additional distribution models so that you can better understand what is in the market today, which you may or may not see as your competition.

Here is a look at the new model:

Direct-to-consumer (DTC) Digital Channels

- **Insurer Websites and Apps:**

  - Customers can directly purchase and manage policies without any human interaction.

- **AI-Powered Chatbots:**

  - Insurers use chatbots to provide instant support, answer questions, and guide users through the policy selection process.

- **Customizable Policies:**

  - Platforms allow consumers to tailor coverage to their specific needs in real-time.

Embedded Insurance

- **Seamless Integration:**

  - Insurance is offered as part of another transaction, removing the need for a standalone purchase process.

- **Examples:**

  - Travel insurance is included during flight booking.

  - Cyber insurance bundled with tech subscriptions.

  - Extended warranties integrated into e-commerce checkout flows.

Online Marketplaces and Aggregators

- **Comparison Tools:**

  - Aggregators like Bold Penguin or Policygenius allow users to compare policies from multiple insurers side-by-side.

  - API Marketplaces

- **Convenience:**

  - Consumers can choose and purchase a policy directly from the aggregator, with clear visibility into pricing and features.

- **Broker Partnerships:**

  - Many aggregators partner with brokers to handle complex or personalized cases.

Partner Ecosystems and Affiliates

- **Partnerships with Non-Insurance Companies:**

  - Insurance is distributed through partnerships with tech firms, retailers, or service providers.
  - Examples:
    - Home insurance is offered by real estate platforms.
    - Health insurance is sold through fitness apps or gym memberships.

- **Affinity Programs:**

  - Insurance products are marketed through associations, clubs, or employer benefits.

Usage-Based and On-Demand Insurance

- **Pay-As-You-Go Policies:**

  - Usage-based auto insurance (e.g., pay-per-mile) and on-demand policies (e.g., hourly or event-specific insurance) are gaining popularity.

- **IoT and Telematics Integration:**

  - Devices like smart home systems or vehicle telematics track usage or behavior, enabling dynamic pricing and personalized coverage.

Microinsurance and Niche Products

- **Focused Coverage:**

  - Small-scale or event-specific policies aimed at underserved markets (e.g., gig workers, freelancers, low-income populations).

- **Examples:**

  - Coverage for a single trip, event, or item (e.g., smartphone insurance).

Digital Agents & Brokers

- **Tech-Enhanced Services:**

  - Brokers are leveraging digital tools to automate routine tasks and focus on strategic advisory services.

- **Agency Platforms:** More agency owners and brokers are developing agency management systems that do more than the typical CRM. These systems may include a quoting tool, leverage AI to replace field underwriting, speed up manual processes, and compare quotes of multiple insurers, all within their platform, to help provide a more seamless and personalized experience for their customers.

- **Hybrid Models:**

  - A mix of digital self-service and human interaction appeals to consumers who want convenience and personal advice.

Proactive Risk Management

- **Wellness Programs and Incentives:**

  - Insurers partner with fitness trackers, telemedicine, or wellness apps to engage with customers and reduce risks.

- **IoT Integration:**

  - Smart devices like water leak sensors or wearable health monitors help insurers proactively reduce claims.

Blockchain and Smart Contracts

- **Secure Transactions:**

  - Blockchain ensures transparency and security in policy issuance and claims processes.

- **Automated Payouts:**

  - Smart contracts trigger automatic payments for pre-defined events (e.g., flight delay insurance).

Social and Peer-to-Peer Models

- **Community-Based Insurance:**

  - Policyholders pool their resources with payouts managed by the community or a digital platform.

- **Crowdsourced Risk Sharing:**

  - Platforms like Lemonade use peer-to-peer models, where unused premiums are donated or returned.

Distribution Model Summary

Channel Type	Description	Example
Agents & Brokers	Personalized advice and policy customization, enhanced by digital tools	Hybrid advisory + CRM platforms
Direct-to-Consumer	Policies are sold directly through websites, apps, and chatbots	Lemonade, Progressive online platforms
Embedded Insurance	Insurance bundled with other products or services	Travel insurance with light bookings
Aggregators	Online comparison tools aggregating multiple insurers	Policygenius, Compare.com
Partner Ecosystems	Insurance is distributed through non-traditional partners like tech firms and retailers	AppleCare, Tesla insurance
Usage-Based & On-Demand	Flexible policies tailored to usage or short-term needs	Pay-per-mile car insurance
Digital Broker	Tech-driven brokers are offering a mix of automation and personal service	Digital-first brokerages like Embroker
Microinsurance	Small, specific policies for underserved or niche markets	Event insurance, gig worker policies
Proactive Risk Management	Insurers are engaging in prevention and real-time risk monitoring via IoT and partnerships	Smart home devices, fitness trackers
Blockchain/Smart Contracts	Transparent and automated claims and underwriting processes	Flight-delay claims automation
Peer-to-Peer	Community-based risk sharing models	Lemonade's giveback program

The insurance industry is transforming to keep up with shifting customer preferences and economic challenges. At the heart of this evolution is a commitment to blending the best of traditional practices with modern digital tools to better serve customers. The critical part of this transformation is agents and brokers whose role is preserved and enhanced. By improving systems like quoting and submissions, insurers are making it easier for them to do their jobs efficiently and focus on what matters most, providing personalized solutions that meet the unique needs of every customer. It's no longer about just selling products—it's about innovatively creating meaningful customer experiences.

This new approach offers agents and brokers significant advantages. With advanced tools and support, they can meet customers wherever they are—online, in person, or through a combination of both. They can tailor products to fit individual needs, moving beyond the one-size-fits-all policies. The future of insurance distribution is a hybrid model that combines the strengths of traditional agents and brokers with cutting-edge technology. This partnership creates an ecosystem where human connection and digital innovation work hand-in-hand. Agents and brokers remain indispensable in delivering the human touch that builds trust, offering guidance and reassurance in complex decisions.

Whether through direct digital sales, embedded insurance, or innovative solutions like microinsurance, the goal is to create a system that meets customers where they are and serves their evolving needs.

TAKE A MOMENT TO REFLECT

How familiar were you with the different distribution models in the insurance industry before reading this chapter? Write about what you knew, what surprised you, and how your perspective has changed.

....

....

....

....

....

....

....

....

....

....

....

....

....

....

....

....

....

TECHNOLOGY INTEGRATIONS TRANSFORMING THE ROLE OF AGENTS AND BROKERS IN PROPERTY & CASUALTY INSURANCE

My motto in business has always been to go wherever my license will take me. That means if today it's personal lines and tomorrow it's cyber insurance, that's what I will do. You must keep continuing to evolve as the landscape changes in the insurance industry. You only need one license to sell both of those products anyway. If that's what my clients are demanding, then that's what my agency is selling. Follow the demand: being the only person in your city offering something in high demand is a great way to position yourself as the go-to person and thought leader for that product.

When I transitioned from being an agent to a broker, I realized how siloed the distribution chain was in insurance. My mentors, who had been in the industry for almost 30 years, rarely understood how agents do business versus a broker, along with the total offering of insurance products and verticals in the market that one could sell with a single license, so they could diversify their book when needed. We all just kind of stayed in our silos and kept our heads down.

I once had the opportunity to sit and meet with the six other leaders when I was an ERG leader. Julio Portalatin, the Former CEO of Mercer, who later became the Vice Chair of Marsh and McLennan

Company, gave tremendous advice that stuck with me forever. I asked him how he prepared himself to be CEO of a Big Four firm. He said he'd worked in every part of the insurance industry and multiple roles to learn how everyone does their jobs and effectively leave them when he became a CEO one day because he had been in their shoes before becoming their leader. I applied that mantra to my Z career path in the insurance and insurtech industries; I have had the opportunity to sell or market insurance products in property & casualty, life & health, and financial products. It gave me an incredible perspective to view opportunities for technology to be integrated into the insurance sector once I entered the insurtech space.

When most people think of innovation in the insurance sector, we think of P&C first. Mainly because we have seen the most innovation in this area over the last 10 years since the entrance of Insurtech. However, there has still been a lot of innovation in the Life and Health sector. We will discuss the trends that we have seen thus far, along with the opportunities that exist in the verticals for agents and brokers to embrace this technology best to help grow your book of business and stay ahead of the game.

Let's discuss the variations of property and casualty & life and health products. We will discuss the product types, changes, opportunities, and trends for innovation in both insurance verticals. This will help you understand what verticals you may currently be working in and whether they are already trending toward innovation. We will also discuss how they may have developed already and how you can get equipped to change quickly; these trends are emerging swiftly.

For those brand new to the industry or thinking of diversifying their focus, here is a breakdown of the industry's core verticals, along with some examples of how agents and brokers can leverage emerging technology in these areas.

Let's start with Property and Casualty (P&C) insurance. It covers various verticals that agents and brokers can specialize in or offer as part of a comprehensive book of business.

Personal Lines Insurance

Homeowners Insurance	Covers damages or losses to residential property and liabilities for injuries occurring on the property
Renters Insurance	Protects tenants' personal belongings and liabilities in rented properties
Condo Insurance	Specifically for condominium owners, covering their unit and personal property
Auto Insurance	Covers physical damage, theft, and liability for vehicles owned by individuals
Personal Umbrella Insurance	Provides additional liability coverage beyond standard policies' limits
Recreational Vehicle Insurance	For boats, motorcycles, RVs, ATVs, and other personal-use vehicles

Let's examine how technology integrations affect how agents and brokers deliver personal lines insurance products.

Homeowners Insurance

Smart home tech like sensors, drones, and IoT devices are changing the game for homeowners insurance. These tools monitor for issues like water leaks, fire risks, or security breaches, helping clients prevent claims before they happen. For agents, it's an opportunity to offer discounts and position yourself as a proactive partner in home protection.

Renters Insurance

Renters love convenience, and digital tools like apps make buying and managing policies more effortless than ever. Smart locks and security cameras not only boost safety but also lower risks. As an agent, highlighting these benefits lets you connect with today's tech-savvy renters.

Condo Insurance

Condo insurance can get tricky, balancing personal coverage with association policies. Digital tools simplify the process, helping clients identify gaps in coverage. These platforms save time and build trust for agents by making tailored recommendations easy.

Auto Insurance

Telematics is transforming auto insurance, using driving data to reward safe habits with lower premiums. For agents and brokers, promoting usage-based policies helps clients save money and sets them apart. Plus, tech-driven claims like virtual inspections speed up resolutions and improve satisfaction, which is helpful for renewal season.

Personal Umbrella Insurance

Digital calculators and risk tools make umbrella policies easier to sell by clearly showing uncovered claims' costs. These tools help agents turn a complex conversation into a simple, compelling case for extra protection.

Recreational Vehicle Insurance

GPS tracking and IoT devices are helping recreational vehicle owners protect their investments. For agents, this means offering more tailored coverage options and giving clients peace of mind, whether for their RV, boat, or ATV.

With technology paving the way for more innovative, more personalized solutions, these tools help agents stand out as forward-thinking, trusted advisors. Embrace them, and you'll do more than keep up—you'll lead the way.

Commercial Lines Insurance

Commercial Property Insurance	Protects business properties such as offices, warehouses, and retail stores from damages or losses
General Liability Insurance	Covers businesses against bodily injury, property damage, and advertising injury claims
Commercial Auto Insurance	Provides coverage for vehicles used in business operations
Workers' Compensation Insurance	Covers medical expenses and lost wages for injured employees
Business Interruption Insurance	Helps businesses recover lost income due to disruptions like fires or natural disasters
Professional Liability (Errors & Omissions)	Covers professionals against negligence or nadequate performance claims
Product Liability Insurance	Protects businesses against claims related to products causing harm or injury
Cyber Liability Insurance	Covers risks associated with data breaches and cyberattacks
Directors and Officers Insurance (D&O)	Protects company executives from personal losses due to lawsuits against their decisions or actions
Commercial Crime Insurance	Provides protection from financial losses related to business-related crime, including employee theft, forgery, robbery, and electronic crime

Let's examine in more detail how technology integrations affect how commercial insurance products are delivered by agents and brokers.

Commercial Property Insurance

IoT sensors and intelligent systems are revolutionizing property insurance, detecting real-time risks like leaks, temperature changes, or unauthorized access. For agents and brokers, this means offering proactive risk management solutions that lower claims and qualify clients for premium discounts.

General Liability Insurance

AI and predictive analytics now identify liability risks by analyzing claims patterns. Agents and brokers can use these insights to highlight coverage gaps and tailor policies, building stronger client trust.

Commercial Auto Insurance

Telematics and GPS tracking optimize fleet management by monitoring driver behavior and maintenance needs. Agents and Brokers can offer usage-based policies that reward safe driving and lower business costs.

Workers' Compensation Insurance

Wearables and AI-driven safety tools reduce workplace injuries by monitoring posture, fatigue, and hazards. Agents and Brokers promoting these tech-integrated policies help clients create safer work environments and cut claims.

Business Interruption Insurance

AI models predict the financial impact of disruptions, providing precise coverage options. Agents and Brokers can simplify conversations about recovery timelines and emphasize the value of this coverage.

Professional Liability (E&O)

AI improves underwriting by evaluating risks faster, while digital tools help clients maintain compliance. Agents and Brokers can recommend tech-driven resources to minimize lawsuits, enhancing client confidence.

Product Liability Insurance

Blockchain and IoT improve supply chain transparency, making it easier to address risks like recalls. Agents and Brokers can offer targeted protection, aligning coverage with specific product vulnerabilities.

Cyber Liability Insurance

AI and blockchain safeguard sensitive data with threat detection and secure data handling tools. Some stand-alone cyber insurance providers even offer complimentary or low-cost employee cybersecurity training embedded in the product. Others offer discounted services to cybersecurity expert services as well. Agents and Brokers can offer these solutions as a value-add to their clients.

Directors and Officers Insurance (D&O)

Analytics tools assess executive risks by tracking regulatory changes and market trends. Agents and Brokers can position D&O insurance as essential protection, strengthening relationships with leadership clients.

Commercial Crime Insurance

Fraud detection software and AI flag suspicious activity, combating theft and financial crimes. Agents and Brokers can guide clients to policies incorporating these tools, showcasing expertise in asset protection.

Commercial insurance has been a hot commodity for innovation in insurance and has seen the most digital transformation over the last few years. Embrace the technology as value added, not something extra you're forced to tell your clients about. Those complimentary risk management tools could result in significant savings that your clients will thank you for when it adds value to their bottom line.

Specialty Insurance

Inland Marine Insurance	Covers goods in transit and movable property
Builder's Risk Insurance	Protects construction projects from damages during the building process
Surety Bonds	Guarantees contractual obligations will be met
Fiduciary Liability Insurance	Covers risks for those managing employee benefits
Environmental Liability Insurance	Covers claims related to pollution or environmental damage.
Event Insurance	Protects against risks associated with hosting events, such as cancellations or liabilities
Aviation Insurance	Covers aircraft and aviation-related risks
Malpractice Insurance	A specialized form of professional liability coverage that protects professionals, particularly in healthcare and legal fields, against claims of negligence or errors that result in harm to clients or patients

Let's take a deeper look into how the technology integrations affect how specialty insurance products are delivered by agents and brokers, along with the value they bring.

Inland Marine Insurance

GPS tracking, IoT sensors, and blockchain have always ensured that goods in transit are better. Real-time updates on shipment location and condition help prevent risks like delays or damage. For agents and brokers, offering tech-enhanced policies makes you a trusted partner in delivering transparency and efficiency.

Builder's Risk Insurance

Drones and AI are transforming construction risk management by identifying hazards and analyzing timelines. By promoting these tools, agents and brokers can help clients protect their projects and stand out in the market.

Surety Bonds

Blockchain simplifies bonding with secure, transparent systems that speed up approvals and reduce paperwork. Agents and Brokers offering these solutions can provide faster, more reliable services, solidifying client trust.

Fiduciary Liability Insurance

AI-powered compliance tools help fiduciaries avoid costly errors and stay on top of regulations. Recommending these technologies shows clients' agents and brokers understand their needs and can offer proactive protection.

Environmental Liability Insurance

IoT sensors and satellite imagery monitor environmental risks, while AI predicts future liabilities. By offering policies backed by these tools, agents, and brokers can help clients mitigate risks and support sustainability efforts.

Event Insurance

Event management software and AI predict disruptions, making large-scale events safer and more manageable. For agents and brokers, offering these tech-driven policies simplifies the process and positions you as a go-to resource.

Aviation Insurance

Telematics and AI optimize aircraft safety and maintenance, reducing risks. Agents and brokers using these tools can offer precise coverage while showcasing expertise in the aviation industry's unique challenges.

Malpractice Insurance

AI improves risk assessment and pricing, while telemedicine brings new liability concerns. Agents and brokers addressing these emerging challenges can provide tailored solutions to keep clients fully protected.

Technology simplifies specialty insurance, allowing agents to deliver tailored, efficient solutions. Whether it's builder's risk, environmental coverage, or event insurance, embracing these tools lets you stand out as a forward-thinking partner. You can lead the way in this evolving field with the right technologies.

Agricultural Insurance

Farm and Ranch Insurance	Covers properties, equipment, and livestock for farms and ranches
Crop Insurance	Protects against losses due to natural disasters or declining crop prices

Let's examine in more detail how technology integrations affect how agricultural insurance products are delivered by agents and brokers and the value they bring. The agricultural insurance industry is evolving faster than ever, driven by the adoption of innovative technologies transforming how we support farmers and manage risks. For agents and brokers, these tools aren't just buzzwords—they're real solutions that make a difference for farmers, governments, and insurers alike. Let's explore how these technologies are shaping the future of agri-insurance and why now is the time to embrace them.

- **Farm Management Software**

  Farmers are using software to track crop health, manage resources, and boost yields—all with just a few clicks. This data is gold. It enables more accurate risk assessment, policy customization, and streamlined claims. As an agent or broker, incorporating these insights into your offerings shows clients that you're not just selling policies—you're providing more innovative, tailored solutions that address their unique needs.

- **Smart Irrigation Systems**

  Smart irrigation ensures crops get the needed water, reducing waste and lowering the risk of drought-related losses. Promoting policies with perks for using smart irrigation helps your clients and positions agents and brokers as forward-thinking partners in agricultural success.

- **Drones and Robots**

  Drones are becoming essential tools for farming, offering quick, cost-effective ways to monitor crops, assess damage, and even speed up claims after storms or disasters. Robots are making farm operations more efficient by handling planting, harvesting, and other labor-intensive tasks. These technologies simplify claims processing and risk evaluation for agents and brokers, making policies more effective and reliable.

- **Precision Agriculture and Predictive Analytics**

  Precision agriculture uses GPS, IoT devices, and data analytics to maximize efficiency, while predictive analytics leverages historical and real-time data to forecast yields, weather, and pest risks. For agents and brokers, this means proactively identifying potential risks and offering dynamic, adaptive policies.

- **Sensors**

  Sensors in fields, equipment, and even livestock provide real-time data on soil moisture, crop health, and more. For agents and brokers, promoting policies integrating these technologies makes you a valuable resource for clients seeking more innovative solutions.

- **Integrated Marketplace Platforms**

  These platforms connect growers, suppliers, buyers, and insurers, simplifying every step of the agricultural process. Farmers see this model as a one-stop shop for everything from resources to insurance. It's a direct way for agents and brokers to offer tailored, bundled services that meet clients' specific needs while deepening relationships in the agri-community.

- **A Digital-Agri Ecosystem**

  The real revolution lies in creating a connected, digital agri ecosystem. This isn't about adopting a single tool—it's about leveraging a network of technologies to go beyond risk coverage and become a true partner in agricultural success. Agents and Brokers who embrace this shift have the opportunity to lead the industry by offering innovative, value-driven solutions.

The future of agricultural insurance isn't just about policies—it's about creating relationships, leveraging new tools, and solving real problems for your clients. This transformation is happening now, and the agents, brokers, and carriers who adapt will not only thrive but set the standard for what agri-insurance can be. This is your moment to lead.

High-Net-Worth Insurance

High-Value Home Insurance	Specialized coverage for luxury homes
Fine Art and Collectibles Insurance	Covers high-value personal property
Private Client Auto Insurance	Tailored coverage for luxury or collector vehicles

Emerging technologies are revolutionizing how we approach high-net-worth insurance products, allowing agents and brokers to deliver tailored, efficient solutions for affluent clients. Let's explore how these innovations shape the landscape for luxury-focused insurance products and what they mean for your agency or brokerage.

High-Value Home Insurance

Luxury homes need more than standard coverage, and smart tech like IoT sensors and security systems are stepping up. Water leak detectors and real-time monitors help prevent costly damage. Agents and Brokers offering these solutions show clients you're protecting their investments, not just insuring their homes.

Fine Art and Collectibles Insurance

Insuring fine art goes beyond valuations—blockchain tracks provenance, while sensors ensure preservation by monitoring humidity and temperature. By offering these advanced tools, you as agents and brokers, reinforce your role as a trusted advisor to clients who treasure their collections.

Private Client Auto Insurance

Luxury cars deserve tailored coverage. Telematics and GPS protect against theft and damage, while AI personalizes premiums. Agents and Brokers, offering these tech-driven solutions help deliver the exceptional service that high-net-worth clients expect.

Technology isn't a bonus for affluent clients—it's an expectation. By using IoT, blockchain, and AI, you can offer bespoke solutions, deepen client relationships, and lead in this exclusive market.

Public Sector and Nonprofit Insurance

Municipality Insurance	Covers risks specific to government entities
Nonprofit Insurance	Tailored for organizations with limited budgets and unique liabilities
School Insurance	Covers educational institutions for property, liability, and unique risks

- **Municipality Insurance**

  Municipalities face unique risks like property damage and cyber threats, and tools like GIS and AI-driven platforms make risk assessments more accurate. For agents, these technologies mean tailored coverage and a clear competitive edge.

- **Nonprofit Insurance**

    Nonprofits need affordable, precise coverage and automated risk assessment tools to help identify vulnerabilities without overpaying. For agents, it's more than selling policies—it's about supporting meaningful causes.

- **School Insurance**

    Schools face risks from property damage to cyberattacks, and smart security systems and real-time monitoring help mitigate them. Agents can position themselves as trusted advisors by offering proactive, tech-driven solutions.

Across these sectors, technology isn't just a tool—it's a game changer. For agents and brokers, it isn't just about paperwork and premiums; it's about technology-powered partnerships, which means delivering value through expertise and innovation.

Industry-Specific Insurance

Hospitality Insurance	For hotels, restaurants, and bars
Retail Insurance	Coverage for stores and retail operations
Construction Insurance	Comprehensive coverage for contractors and construction firms
Technology Insurance	Covers risks specific to tech companies, including intellectual property
Transportation Insurance	Coverage for trucking, logistics, and other transport-related businesses
Energy Insurance	Tailored for oil, gas, and renewable energy industries
Healthcare Insurance	Protects medical practices and facilities

- **Hospitality Insurance**

  Smart technology like IoT sensors helps hospitality businesses detect risks like fires and track occupancy, reducing claims and improving pricing for clients. This makes agents and brokers trusted risk management partners.

- **Retail Insurance**

  Tech tools like inventory tracking and point-of-sale integration allow insurers to offer adaptive policies, helping agents and brokers deliver tailored solutions for evolving retail risks.

- **Construction Insurance**

  Drones and wearables improve safety and reduce claims on construction sites, enabling agents and brokers to offer better coverage and keep premiums manageable.

- **Technology Insurance**

  Cybersecurity tools now assess risks like data breaches and IP theft, empowering agents and brokers to provide tailored coverage and guide clients in safeguarding their innovations.

- **Transportation Insurance**

  Telematics tracks driver behavior and vehicle health, reducing accidents and costs while helping agents and brokers offer safer, more efficient policies.

- **Energy Insurance**

  Predictive analytics and sensors monitor equipment to prevent failures, allowing agents and brokers to craft specialized policies and reduce claims.

- **Healthcare Insurance**

  EHR and telemedicine tools enhance risk assessment for malpractice and data breaches, allowing agents and brokers to offer precise, customized healthcare coverage.

Thanks to technology, insurance is no longer just about reacting to risks but proactively managing them. By embracing these emerging technologies, agents and brokers can deliver more responsive and customized insurance solutions, effectively addressing each industry's unique challenges. This is more than insurance—it's a partnership for progress.

Catastrophic and Disaster Insurance

Flood Insurance	Covers damages caused by flooding
Earthquake Insurance	Provides coverage for earthquake-related damages
Hurricane and Windstorm Insurance	Tailored for regions prone to such disasters
Parametric Insurance	Pays a fixed amount when predefined conditions or triggering events are met rather than compensating for actual losses

Flood Insurance: High Waters, Smarter Solutions

Flooding is one of the most devastating disasters, and predicting its impact used to be a guessing game. Not anymore.

Advanced mapping technologies and satellite imaging now provide real-time flood risk assessments, helping insurers price policies more accurately. For brokers and agents, this means fewer surprises for clients and the ability to offer coverage tailored to specific risk levels; this also helps them build trust by assisting clients to protect their homes and businesses proactively.

Earthquake Insurance: Seismic Shifts in Coverage

Earthquakes strike with little warning, but technology is changing the game. Seismic monitoring systems and AI-driven risk models allow insurers to predict potential damage zones more accurately. For agents and brokers, this means offering highly customized policies based on geographic data and building materials and leveraging data-backed insights that make it easier to explain coverage options to clients.

Hurricane and Windstorm Insurance: Weathering the Storm with Tech

Hurricanes and windstorms wreak havoc, so predictive analytics and real-time weather tracking help insurers adjust coverage and premiums based on evolving storm patterns. Drones and IoT sensors assess property vulnerabilities before and after storms, streamlining claims processes. Brokers and agents can offer proactive strategies to minimize risk and quickly assist clients in recovering from damages.

Parametric Insurance: Predictable Payouts for Unpredictable Events

Parametric insurance relies heavily on technology, such as satellite data and IoT devices, to monitor triggering events. For agents and brokers, it's a unique product to offer clients who value speed and simplicity. It also eliminates the hassle of lengthy claims processes, and clients appreciate getting funds quickly when disaster strikes.

Technology is making disaster-related insurance smarter, faster, and more responsive. For agents and brokers, this is an opportunity to be heroes for their clients—offering solutions that don't just cover risks but actively help manage and reduce them.

REFLECTION

Scenario 1: Using Smart Technology in Personal Lines Insurance

You're meeting with a homeowner who wants to switch insurance providers. They are interested in finding ways to protect their property from common risks. What specific technology-driven solutions would you highlight to demonstrate that your offering goes beyond traditional coverage and shows the value of your policy?

....

....

....

....

Scenario 2: Leveraging Predictive Analytics for a Commercial Property Insurance

A business owner is concerned about protecting their retail store from potential risks like water damage, fire, or theft. They're also looking for ways to reduce premiums while maintaining comprehensive coverage. What technologies would you highlight to demonstrate the value of your commercial property insurance policy? How would you position these tools as proactive risk management solutions to build trust?

....

....

....

....

CHAPTER **FIVE**

TECHNOLOGY INTEGRATIONS TRANSFORMING THE ROLE OF AGENTS AND BROKERS IN LIFE AND HEALTH INSURANCE

INNOVATION

As we step into the life and health space, let's go beyond the basics and really unpack what's possible. There's so much room for technology to make a difference in areas like life insurance, health coverage, group benefits, Medicare, and retirement products. The opportunities are endless, and I'm here to guide you through how these integrations can transform how we approach these essential products. Let's dig in!

Previously, there hasn't been as much noise around innovation in this space since most of the funding went to P&C products. Life and health (L&H) insurance is holding its own for the first time in three years, attracting 50% of 2024's funding alongside property and casualty (P&C). It's a big deal.

On the L&H side, the health sector is driving this surge, while in P&C, it's all about climate risks and business insurance. This shift shows just how much both sectors are evolving and it's exciting to see funding balancing out between these two critical areas.

We have already seen some carriers who were pretty early adopters of either insurtech partnerships or being a leader in insurance innovation. There are numerous carriers and brokerages with insurance innovation departments who are working on some pretty incredible solutions. Often, that innovation is not meant to be shared outside of their organizations; they are either innovating to create better customer experiences or streamlining their in-house processes and workflows.

There are a few companies that I have seen do well in innovation, whether through partnerships or in-house innovation within these verticals. I admire the work that Life & General America is doing in insurtech partnerships, they were early adopters of insurtechs such as Ehtos Life, Lemonade, and others. The great thing is within a lot of the carrier partnerships, they don't remove agents from the equation, they still make sure that their insurtechs process always touches an agent within the client experience instead of removing them from it. I also enjoyed learning about GRAIL, a biotech company that partnered with brokers and carriers to offer an incredible benefit to their policyholders. The Galleri test is a blood test that screens for DNA of various cancer cells. It is risk management for the life insurance carrier, broker, and client wrapped in one value-added product.

Wellness Innovation

We are seeing more innovation around wellness products, including wearable technology and (IOT the Internet of Things) to offer personalized care and leverage health insurance plans through predictive analytics. By integrating big data and analytics with wearables and wellness apps, you can now provide clients with health and life insurance solutions tailored to their unique needs and lifestyles. Prudential is doing some pretty incredible work in the group benefits technology space. I had the opportunity to have their Director of Benefits Technology guest lecture for my UCONN

Ms Fintech class, where I'm also an Adjunct Professor. They learned so much from that lecture, they are leveraging technology in some pretty unique ways. Prudential introduced a digital well-being hub—a centralized, user-friendly platform that combines all their life, absence, and disability products. This innovative tool acts as a one-stop shop, making it easier for employees and members to access the resources they need when they need them, by simplifying access and streamlining their offerings. It's a forward-thinking approach that enhances convenience and underscores the importance of holistic care in today's benefits landscape.

RetireTech & WealthTech

Let's explore two exciting areas reshaping the retirement landscape: retiretech and wealth tech. These spaces are brimming with innovations that benefit clients and create incredible opportunities for agents and brokers to expand their offerings and add real value.

Retiretech is all about simplifying and modernizing retirement planning. Solutions like digital retirement calculators, automated savings platforms, and tools for managing pensions or 401(k)s are making it easier than ever for clients to plan for their futures. For agents and brokers, these tools provide a way to engage clients in deeper conversations about long-term planning, offering them tailored options that align with their goals. It's a chance to move beyond traditional products and position yourself for life's biggest milestones.

Wealthtech, on the other hand, focuses on growing and managing wealth. From AI advisors and investment platforms to financial wellness apps, this space is all about making financial tools accessible and efficient. Wealthtech opens the door to cross-selling opportunities for agents and brokers, integrating investment solutions with existing insurance products. Imagine helping clients plan generational wealth or introducing them to cutting-edge tools like cryptocurrency investment platforms—suddenly, you're not just their insurance expert, but their go-to financial resource.

These tools not only make your work easier but also help you build stronger, more holistic relationships with your clients.

A few upcoming players are creating some revolutionary solutions to help retirees who are still working full-time positions and staying on the traditional group benefits plans. The retirement market is making way for more tech-forward retirement planning platforms for advisors and financial representatives to help them better model annuities and retirement models for their clients. Even the approach to leveraging annuities has changed, more agents and brokers are encouraging generational wealth transfer. Great opportunities exist to leverage technology in these areas.

Artificial intelligence (AI)

Generative AI and AI-powered chatbots are flipping the script on how we connect with clients, making communication faster, smoother, and incredibly personalized. But let me get real with you—this isn't just about technology, it's about meeting the needs of today's reality. Health insurance carriers are teaming up with wellness companies, mental health platforms, and caregiving support systems to address a growing challenge: the "sandwich generation" of millennials. Yep, I'm part of that crew—managing toddlers, teenagers, aging parents, and grandparents, all while running a global insurtech consulting firm. Sound familiar?

This is where agents and brokers can shine by leveraging these emerging technologies to add client value. AI doesn't just streamline processes like trend analysis or policy renewals; it enables personalized solutions that resonate with what clients need in real-time. Whether working in life, health, P&C, or any other vertical, these tools can help you stand out by delivering smarter, more intuitive services. This isn't just about keeping up—it's about leading the charge in making insurance work for today's complex lives.

I'm thrilled to see how Generative AI, Conversational AI, and similar technologies are making waves in our industry. One area catching my attention is the rise of conversational AI call center solutions in the senior benefits. These tools do so much more than just handling FAQ calls—they're stepping in during high-pressure Medicare enrollment seasons to ensure no sales lead is missed.

What's exciting is that these solutions aren't replacing agents; they're amplifying their effectiveness. Most systems I've come across act as appointment setters, seamlessly lining up qualified leads for agents to close deals faster—even after hours. It's like having a full agency team working around the clock during open enrollment, without the added overhead. This means more time for agents to focus on what they do best: building relationships and delivering solutions. It's a powerful reminder of how tech can enhance, not replace, the human touch in this business.

Health Innovation

Since the pandemic, millions of people have turned to telemedicine, and honestly, it's been a game-changer for health insurance carriers. Pairing telemedicine with wearables takes it to the next level. IoT and wearable devices allow policies to be tailored to real-time health data, which means better client satisfaction, smarter risk management for insurers, and lower treatment costs through preventive care. It's a win all around.

Blockchain steps in to ensure that transactions and health records are secure and tamper-proof, which not only streamlines the process but also builds trust. For agents and brokers, this opens the door to tools that help create personalized health and wellness plans, allowing you to deliver value in ways that truly resonate with today's tech-savvy clients. By staying ahead of these trends, you're not just selling policies—you're offering proactive solutions that empower clients to manage their health risks while helping insurers reduce claims. Everyone benefits. So, think about partnering with service providers in these areas to offer more than just coverage; offer value-added services that set you apart and strengthen the relationships you've worked so hard to build.

Let's dive into life and health (L&H) insurance. These products cover a wide range of verticals, each offering unique opportunities for agents and brokers to specialize or build a well-rounded book of business. Whether you're new to the industry or considering expanding your focus, understanding these areas is key to staying ahead.

We'll explore the types of products available, the changes and trends shaping the market, and the exciting opportunities for innovation. This isn't just about keeping up—it's about being prepared to pivot quickly as these trends emerge and evolve. Together, we'll look at how some of these verticals are already adapting, and I'll share ways you can leverage technology to not only meet client expectations but exceed them. Whether deep into L&H or just starting to explore, this breakdown will give you the insights you need to thrive in this space.

As someone in the insurance and insurtech world for 15 years, I've seen firsthand how embracing innovation can elevate your business and strengthen your relationships with clients. Here's how technology shapes each type of life, health, and benefits insurance and how you can leverage it to stand out as an agent or broker.

Life, Health, and Benefits:

Life Insurance

Term Life Insurance	Provides coverage for a specific period, with a payout only if the insured dies within the term
Whole Life Insurance	Permanent coverage with a cash value component that builds over time
Universal Life Insurance	Flexible coverage with adjustable premiums and savings options
Variable Life Insurance	Combines life coverage with investment options tied to market performance
Final Expense Insurance	Simplified life insurance designed to cover end-of-life expenses like funerals
Group Life Insurance	Offered to employees or members of an organization as part of a benefits package
Key Person Insurance	Protects businesses by compensating for financial losses caused by the death of a key employee or executive

- **Term Life Insurance**

  Tech tools like AI-powered underwriting and predictive analytics are speeding up the approval process, often issuing policies within hours. For agents and brokers, this means fewer roadblocks and faster sales. Clients love the convenience—so why not highlight this seamless experience to close more deals?

- **Whole Life Insurance**

  With digital policy management platforms, clients can easily track their cash value growth and manage premiums. This transparency builds trust and keeps clients engaged long-term. Positioning these features as part of your value as an agent or broker can set you apart from the competition.

- **Universal Life Insurance**

  Smart financial calculators and interactive dashboards allow clients to adjust premiums and savings as their needs change. This flexibility is a big selling point, and as an agent or broker, you can show clients how these tools help them stay in control of their financial future.

- **Variable Life Insurance**

  Clients who want their insurance policies tied to market performance can now monitor investments in real-time through user-friendly platforms. Agents and brokers can use this tech to give clients the confidence to align their coverage with broader financial strategies.

- **Final Expense Insurance**

  Simplified underwriting and online applications make it easy for clients to get coverage quickly. For agents, this means targeting clients who value no-hassle solutions while offering a product that meets a critical need.

- **Group Life Insurance**

  HR integration tools and digital enrollment platforms take the headache out of managing group life policies. Employers appreciate how easy these systems make benefits administration, giving brokers and agents a strong value proposition to bring to the table.

- **Key Person Insurance**

  Risk modeling software allows businesses to quantify the financial impact of losing a key team member. As an agent or broker, you can use these tools to deliver tailored solutions that protect businesses and demonstrate the real value of key person insurance.

Life insurance isn't just about policies anymore—it's about personalization, speed, and efficiency. Technology gives you the tools to deliver all that while strengthening your client relationships. Embrace these innovations, and you'll position yourself as a leader in this ever-evolving industry.

Health Insurance

Individual Health Insurance	Coverage purchased by individuals or families, often through marketplaces or private insurers
Group Health Insurance	Employer-provided coverage for employees and their families
High-Deductible Health Plans (HDHPs)	Lower premiums with higher deductibles, often paired with Health Savings Accounts (HSAs)
Medicare Insurance	
Original Medicare (Parts A and B)	Government-provided coverage for hospitalization and medical services
Medicare Advantage (Part C)	Comprehensive plans offered by private insurers, often including additional benefits
Medicare Supplement (Medigap)	Covers costs not included in Original Medicare
Medicare Part D	Prescription drug coverage

Short-Term Health Insurance	Temporary coverage for individuals during gaps between health plans
Critical Illness Insurance	Provides a lump sum for serious illnesses like cancer, heart attack, or stroke
Long-Term Care Insurance	Covers extended care services like nursing homes, in-home care, or assisted living

Disability Insurance	
Short-Term Disability	Covers lost wages for temporary disabilities
Long-Term Disability	Provides income protection for longer-lasting disabilities

- **Individual Health Insurance**

  Tech like online marketplaces and recommendation engines is helping clients compare plans that fit their unique needs. As an agent, you can use these tools to guide clients through a maze of options, positioning yourself as their go-to resource for clarity and expertise.

- **Group Health Insurance**

  Benefits administration platforms are streamlining enrollment and plan management for employers. For brokers, this means you can present group health as an easy-to-implement solution that reduces stress for employers while providing great coverage for their teams.

- **High-Deductible Health Plans (HDHPs)**

  Apps designed for managing Health Savings Accounts (HSAs) give clients real-time insights into contributions and spending. As a broker, showcasing these tools can help you explain the advantages of HDHPs, making clients feel confident about managing out-of-pocket costs.

- **Medicare Insurance**

- **Original Medicare (Parts A and B):** Enrollment platforms and eligibility checkers simplify the process for clients, while you, as their agent, can cut through the red tape to make the experience smoother.

- **Medicare Advantage (Part C):** Predictive analytics help tailor coverage to each client's unique health and lifestyle needs. You can position these plans as an all-in-one solution with added benefits that seniors will love.

- **Medicare Supplement (Medigap):** Online comparison tools show clients how Medigap fills the gaps in their Medicare coverage. Use these tools to make the value of supplemental policies crystal clear.

- **Medicare Part D:** Prescription cost calculators guide clients to the most affordable plans for their medications. Helping seniors navigate Part D can set you apart as someone who truly understands their needs.

- **Short-Term Health Insurance**

  Simplified applications and instant approvals make temporary coverage a breeze. For clients in a pinch, you can be their go-to for quick, stress-free solutions.

- **Critical Illness Insurance**

  AI-driven calculators analyze personal health histories to estimate coverage needs. You can leverage this technology to offer highly personalized policies that give clients peace of mind against major health events.

- **Long-Term Care Insurance**

  IoT devices and wearables track health data to predict future care needs. This gives you a unique opportunity to recommend policies that align with clients' evolving requirements, showing them you're thinking ahead.

Disability Insurance

- **Short-Term Disability:** Automated claims processing tools make sure clients get support when they need it most. Highlighting these tools can show clients how quickly they'll get the help they need.

- **Long-Term Disability:** Financial planning apps integrate disability insurance into broader income protection strategies. Use these tools to demonstrate how this coverage protects clients' futures, adding depth to your offerings.

Technology is transforming the health insurance space, but here's the thing—it's not replacing the human touch; it's enhancing it. By embracing these innovations, you're not just selling policies; you're solving problems, building trust, and delivering personalized solutions that meet the moment. That's the kind of value that keeps clients coming back.

Dental and Vision Insurance

Individual Dental Plans	Covers routine dental care and treatments
Group Dental Plans	Employer-provided dental benefits
Vision Insurance	Covers eye exams, glasses, and contact lenses for individuals and groups

- **Individual Dental Plans**

  Online enrollment tools and automated renewals make signing up for coverage quick and painless. For agents, this means providing clients with a seamless, no-hassle experience while simplifying your workload. Use these tools to show clients that routine dental care doesn't have to be complicated.

- **Group Dental Plans**

  Benefits management platforms allow employers to offer dental coverage while keeping everything organized for their teams. Brokers and agents can position group dental plans as a high-value, low-effort benefit for employers and their employees. Highlight these tools to demonstrate how easy it is to implement and manage coverage.

- **Vision Insurance**

  Digital comparison tools and calculators help clients find the best coverage for eye exams, glasses, and contacts. Brokers and Agents can confidently guide individuals and groups through their options, making the process simple and stress-free. Using these tools, you position yourself as a modern, client-focused professional who puts convenience first.

Dental and vision insurance may seem like the basics, but trust me—there's a lot of opportunity here to elevate the client experience with technology. I've learned that even the smallest tech integrations can significantly impact. These tools make life easier for clients and help agents and brokers stand out as trusted, tech-savvy advisors.

Employee Benefits Insurance

Voluntary Benefits	Optional benefits like accident insurance, hospital indemnity, and critical illness coverage that employees can choose
Flexible Spending Accounts (FSAs)	Tax-advantaged accounts for healthcare and dependent care expenses
Health Savings Accounts (HSAs)	Tax-advantaged savings accounts tied to high-deductible health plans
Wellness Programs	Incentives for healthy behaviors, often including gym memberships, smoking cessation, or health screenings
Employee Assistance Programs (EAPs)	Resources for mental health, financial counseling, and crisis management

- **Voluntary Benefits**

  Online enrollment platforms and personalized recommendation engines guide employees to select optional benefits that fit their needs. Agents and brokers can offer tailored solutions to employers, making voluntary benefits a key part of a comprehensive package. Use these tools to boost employee participation and demonstrate value to your clients.

- **Flexible Spending Accounts (FSAs)**

  Mobile apps and digital dashboards make it easy for employees to track and use their FSA funds in real time. Brokers can position FSA's as a simple, tax-saving benefit while providing tools that make employee management effortless. Highlight these features to show employers how this benefit can enhance their offerings.

- **Health Savings Accounts (HSAs)**

  Banking integration and AI-powered calculators help employees maximize savings and plan for medical expenses. Brokers can educate clients on pairing HSAs with high-deductible health plans to lower costs while empowering employees with tools to manage their healthcare spending. Use these tools to make complex concepts like tax advantages easy to understand.

- **Wellness Programs**

  Platforms that track wellness goals and incentivize healthy behaviors through gamification and rewards programs. Brokers and agents can help employers promote healthier workforces while reducing long-term healthcare costs. Showcase these tools as a way to build employee engagement and strengthen company culture.

- **Employee Assistance Programs (EAPs)**

  Digital portals and telehealth platforms offer employees access to mental health support, financial counseling, and crisis resources. Brokers can emphasize the value of EAPs in addressing employees' holistic well-being while helping employers build a more resilient workforce. Leverage these tools to demonstrate how this benefit supports both productivity and employee satisfaction.

Employee benefits are where insurance meets real-life—helping people protect their health, finances, and well-being in meaningful ways. I've seen how technology is elevating employee benefits, making them more accessible, personalized, and impactful for employees while also simplifying life for agents and brokers.

Supplemental Insurance

Accident Insurance	Covers costs associated with accidents, including medical expenses and recovery
Hospital Indemnity Insurance	Provides a cash payout for hospital stays or specific treatments
Cancer Insurance	Tailored coverage for cancer treatment-related expenses
Travel Insurance	Coverage for medical emergencies while traveling abroad
Pet Insurance	Offers coverage for veterinary expenses for pets, increasingly popular as an employee benefit

- **Accident Insurance**

  Digital claims processing and instant benefit payout platforms help clients get the financial support they need quickly after an accident. Agents and brokers can use these tools to offer fast, stress-free claims experiences, highlighting the value of accident insurance as a reliable safety net. Show clients how these features simplify recovery during a difficult time.

- **Hospital Indemnity Insurance**

  AI-driven calculators estimate the costs of hospital stays, helping clients understand how this coverage can fill financial gaps. Agents can use these tools to demonstrate the real-world benefits of cash payouts for hospitalizations. Highlight this product as a practical, flexible solution for covering out-of-pocket expenses.

- **Cancer Insurance**

  Wellness apps and predictive analytics offer proactive health insights while helping clients manage costs related to treatment. Agents can position cancer insurance as a personalized safety net, providing financial support when it's needed most. Leverage these tools to show clients how this coverage fits into a holistic health plan.

- **Travel Insurance**

  Mobile apps provide real-time assistance for medical emergencies abroad, from finding hospitals to managing claims on the go. Agents can emphasize convenience and peace of mind, showing clients how travel insurance ensures they're protected wherever they go. Highlight tech-enabled features that simplify claims and emergency support.

- **Pet Insurance**

  Online enrollment tools and veterinary expense tracking apps make it easier for pet owners to manage their policies. Agents and Brokerscan present pet insurance as a modern, sought-after benefit for individuals and employees alike. Use these tools to connect with pet-loving clients and showcase how coverage supports their furry family members.

Supplemental insurance isn't just a "nice-to-have"—it's becoming essential for clients looking to fill gaps in traditional coverage. I've seen how technology makes these products easier to understand, sell, and use. For agents and brokers, technology transforms these products into accessible, practical solutions that clients can trust.

Executive and High-Net-Worth Benefits

Executive Benefits	Tailored insurance solutions for executives, such as deferred compensation and supplemental health coverage
Life Insurance trusts	Policies structured to maximize estate planning for high-net-worth individuals
Private Health Plans	Concierge-level health insurance for high-income individuals with access to elite medical networks

- **Executive Benefits**

  Digital platforms manage deferred compensation plans and supplemental health benefits, offering transparency and seamless customization. Agents and brokers can provide executives with tailored solutions that are easy to manage, reinforcing their value as a trusted advisor. Leverage these tools to demonstrate how your offerings protect and enhance an executive's compensation package.

- **Life Insurance Trusts**

  Estate planning software and policy optimization tools help high-net-worth clients structure trusts effectively while maximizing tax benefits. Brokers and agents can use these tools to show clients how life insurance trusts fit into a broader wealth preservation strategy. Position yourself as a partner in safeguarding their legacy with precision and care.

- **Private Health Plans**

  Concierge platforms provide clients with access to elite medical networks and 24/7 telehealth services, ensuring personalized, top-tier care. Brokers and Agents can present these plans as more than just insurance—they're a gateway to exclusive healthcare experiences. Highlight these features to appeal to high-income clients looking for health solutions that align with their lifestyle.

Executive and high-net-worth benefits require a personalized touch, and technology enables you to deliver that effortlessly. By embracing these tools, you can cater to this exclusive clientele with precision and professionalism, setting yourself apart as an advisor who truly understands their unique needs, while providing unparalleled value.

Retirement and Financial Products	
401(k) Plans	Employer-sponsored retirement savings plans
403(b) Plans	Retirement savings plans for non-profits and educational organizations
Annuities	
Fixed Annuities	Guaranteed income for life or a set period
Variable Annuities	Income based on market performance of investments
Indexed Annuities	Returns linked to a market index, like the S&P 500

Deferred Compensation Plans	Programs allowing employees to defer income to a future date
Pension Plans	Employer-sponsored defined benefit plans offering retirement income

Retirement and financial products are key to helping clients secure their futures, build wealth, and plan for the next generation.

401(k) Plans

Digital platforms allow employees to track contributions, manage investments, and project retirement outcomes in real time. Brokers can use these tools to demonstrate the value of employer-sponsored plans while making enrollment seamless. They can also highlight these features to employers to enhance their benefits packages and attract top talent.

403(b) Plans

Automated portfolio management tools and educational apps simplify retirement planning for employees in non-profits and education. Agents and Brokers can position these plans as easy-to-manage solutions for mission-driven organizations. Use tech tools to show how employees can align their investments with retirement goals.

Annuities

- **Fixed Annuities:**

  Online calculators help clients understand guaranteed income projections. Agents can illustrate the stability of fixed annuities, making them a cornerstone of reliable retirement planning. Leverage these tools to simplify conversations about income security.

- **Variable Annuities:**

  Investment tracking apps offer real-time updates on performance. Agents can help clients feel in control by showing how their annuity aligns with market trends. Use these apps to make market-linked annuities more approachable.

- **Indexed Annuities:**

  Tools that track index performance and calculate potential returns provide transparency. Agents can show clients how indexed annuities balance growth potential with security. Position these tools as a way to offer both confidence and opportunity.

Deferred Compensation Plans

Platforms for tracking and customizing income deferral options allow employees to optimize tax benefits. Brokers can present these plans as a smart, flexible tool for long-term wealth building. Use these tools to show executives how deferred compensation aligns with their financial goals.

Pension Plans

Digital pension calculators and plan management tools provide clarity on expected benefits and timelines. Agents can help employers and employees navigate the complexities of defined benefit plans with confidence. Use these tools to position pensions as a valuable part of a comprehensive retirement strategy.

Honorable Mention: Generational Wealth Transfer

Estate planning software and digital financial dashboards allow clients to visualize their wealth transfer plans in detail. Agents can use these tools to help clients structure their financial legacy, ensuring smooth transitions and minimize tax burdens. Highlight your expertise in aligning retirement and financial products with long-term wealth preservation goals.

Retirement and financial products are about more than numbers—they're about helping clients feel secure and confident in their futures. With the right tech tools, you can simplify the process, empower your clients, and position yourself as a trusted partner in building wealth that lasts for generations.

Group Voluntary Benefits

Group Disability Insurance	Short-term and long-term disability options offered to employees
Group Term Life Insurance	Basic life insurance provided as part of a benefits package
Supplemental Group Health Benefits	Additional coverage options such as critical illness or accident insurance
Paid Family Leave Insurance	Offers income protection during family or medical leave

- **Group Disability Insurance**

  Claims automation platforms and predictive risk tools streamline the approval process for short- and long-term disability claims. Agents can highlight faster claim resolutions and seamless support during challenging times. Use these tools to show employers how disability insurance creates a safety net that boosts employee satisfaction and productivity.

- **Group Term Life Insurance**

  Digital enrollment systems and convertible coverage options allow employees to easily manage their policies, even if they change jobs. Brokers can position group life insurance as a hassle-free, foundational benefit. Leverage tech to emphasize convenience and flexibility, making it a no-brainer for employers to include in their offerings.

- **Supplemental Group Health Benefits**

  Personalized recommendation engines and mobile apps help employees select add-ons like critical illness or accident insurance based on their needs. Agents can guide employers in offering targeted supplemental options that enhance core benefits. Highlight these tools to simplify enrollment and boost participation rates among employees.

- **Paid Family Leave Insurance**

  Digital platforms track leave usage, calculate payouts, and integrate with payroll systems for a seamless experience. Brokers can present paid family leave as a modern, essential benefit for today's workforce. Use these tools to showcase how this coverage supports employees during key life events while keeping employers compliant and competitive.

Group voluntary benefits are about more than just adding extras—they're about creating a benefits package that resonates with employees' needs. By leveraging these technologies, agents and brokers can help employers deliver impactful solutions that foster loyalty, productivity, and a positive workplace culture.

Niche Markets

Health and Wellness Programs	Coverage tied to holistic wellness initiatives, often incentivized with IoT devices like fitness trackers
Student Health Insurance	Tailored health plans for college students
Expatriate Health Insurance	Coverage for individuals living or working abroad
Association Health Plans	Group health coverage offered through professional or trade associations

- **Health and Wellness Programs**

  IoT devices like fitness trackers and wellness apps incentivize healthy behaviors by monitoring activity levels, sleep, and overall health metrics. Agents and brokers can offer coverage tied to wellness initiatives, positioning these programs as a way to reduce long-term health risks and premiums. Leverage these tools to show clients how prioritizing wellness benefits their health and finances.

- **Student Health Insurance**

  Online enrollment portals and telehealth services ensure students can easily access and manage their coverage while away at school. Brokers can present student health plans as convenient and affordable solutions for college-bound families. Highlight digital tools to reassure clients that their students can access care no matter where they are.

- **Expatriate (Expat) Health Insurance**

  Global telemedicine platforms and real-time claim-tracking apps provide seamless access to care and benefits abroad. Agents and brokers can position expatriate plans as comprehensive solutions for individuals living or working internationally. Use these tools to emphasize how clients can feel secure knowing they're covered wherever life takes them.

- **Association Health Plans**

  Digital platforms simplify group enrollment and provide members with tools to customize coverage to their needs. Brokers can offer these plans as cost-effective solutions that strengthen member loyalty within trade or professional associations. Use tech tools to streamline enrollment and demonstrate these plans' unique value to members.

  Niche markets like these allow you to serve clients with highly tailored solutions while leveraging technology to provide exceptional service. By embracing these innovations, you'll not only meet your clients' needs—you'll exceed their expectations.

Let's take a moment to reflect on how you can embrace the opportunities we've explored. Use the questions below to think through how technology can elevate your work in life and health insurance, group benefits, niche markets, or retirement planning:

1. **Where do you see the biggest challenges in your current workflow?**

    - Is it time-consuming processes, lack of customer engagement, or inefficiencies in product offerings?

2. **What tools or innovations excite you the most from this chapter?**

    - Are you drawn to AI-powered underwriting, wearable tech for wellness programs, or retiretech platforms? Why?

3. **How can you leverage these technologies to serve your clients better?**

    - Think about the specific needs of your clients. Which tools could address those needs and set you apart from the competition?

....

....

4. **Who are your potential partners?**

-   Whether it's InsurTech providers, carriers, or wellness com-
    panies, who could help you integrate these solutions and
    grow your business?

....

....

....

....

Take a few minutes to jot down your answers. These reflections
can be the starting point for actionable steps toward transforming
your agency or brokerage.

As we close this chapter, let me remind you that technology is reshaping our industry in ways we couldn't have imagined even a de-
cade ago. The opportunities in life and health insurance, group benefits, niche markets, and retirement planning are vast and growing
daily. This shift is exciting because it empowers agents and brokers
to provide more personalized, efficient, and impactful solutions.

This isn't just about keeping up—it's about leading. It's about
leveraging tools and partnerships to build deeper client relationships, address real-life challenges, and position yourself as a trusted advisor in a rapidly changing world. Whether it's integrating IoT
devices into wellness plans, using AI to streamline claims, or exploring the emerging retiretech and wealthtech spaces, these innovations aren't replacing the human touch—they are amplifying it.

So, as you move forward, think big, stay curious, and embrace the
possibilities. This is your moment to transform not just your business but your clients' lives. Let's keep building something incredible together.

CHAPTER **SIX**

DON'T FEAR THE INSURTECH (BALANCING DREAMS AND DISCIPLINE)

Don't let the term 'insurtech' intimidate you—these organizations are not here to disrupt but to protect and elevate the entire insurance ecosystem, ensuring that innovation works hand in hand with the trust and stability this industry has built over generations.

Being in this space for over two decades, specifically in the insurance and technology sectors, has given me a front-row seat to how innovation and regulation intertwine, shaping the industry in inspiring and complex ways. I've come to appreciate the tension between innovation and regulation. It's a dance—sometimes graceful, often clumsy—that shapes not just the industry but the lives of the people it serves. At the heart of this dance group are three pivotal players: the Insurtech Coalition, the American Insurtech Regulatory Council (AIRC), and the National Association of Insurance Commissioners (NAIC). Each brings its own perspective to the table, creating a challenging and necessary dynamic.

Let's start with the Insurtech Coalition. They're the dreamers, the ones lighting up the path forward. Their passion is contagious—championing game-changing technologies like Gen AI, artificial intelligence, machine learning, blockchain, and IoT to reinvent the way insurance works. Imagine a setup where your smart home sensors detect potential issues like carbon monoxide or fire and notify your insurer to send help before it causes major damage. That's their vision in action: turning insurance into a proactive partner that simplifies life.

Then there's the AIRC. If the Coalition is all about dreaming big, the AIRC is the thoughtful voice reminding us to think things through. Their role isn't to pump the brakes on progress but to ensure innovation unfolds in a fair and responsible way. For example, while the Coalition might applaud AI for streamlining claims, the AIRC is asking critical questions: Does this AI treat all customers equitably? Are privacy concerns being addressed? Their oversight ensures that progress doesn't come at the expense of trust.

And finally, the NAIC—the unsung hero working tirelessly behind the scenes like "the Dark Knight of Gotham". This organization ensures consistency and fairness across states, translating bold ideas into practical, actionable policies. While the Insurtech Coalition promotes using drones to assess property damage after a storm, and the AIRC examines privacy concerns, the NAIC develops guidelines that help states adopt these innovations in a way that protects consumers and keeps standards uniform.

Each of these groups brings something vital to the conversation. The Insurtech Coalition drives forward, constantly reimagining what's possible. The AIRC ensures we don't lose sight of ethics and equity along the way. The NAIC pulls it all together, creating the framework that allows these visions to take root without chaos. Together, they're like the parts of an engine, each performing a different but essential role to move us forward.

This kind of collaboration matters deeply. Innovation and accountability aren't just buzzwords to me—they're the building blocks of an insurance industry consumers can trust. Together, they create an environment where technology and regulation enhance each other instead of competing.

When I teach graduate students about the intersection of insurance and technology, I always stress that progress needs both ambition and responsibility. The magic happens when dreamers and watchdogs collaborate, pushing each other to do better. The result?

Solutions that are bold, ethical, and enduring. It's not always a smooth road, but it's one worth traveling to build an industry that truly serves us all—with integrity and ingenuity at its core.

Integrating cutting-edge InsurTech solutions with legacy systems is one of the toughest puzzles insurers, agencies, and brokerages face today. Those legacy systems are like the old-school classics of the industry—solid, dependable, and deeply ingrained in how everything runs. But let's be real: as much as we respect the history, you can't keep trying to tape a smartphone to a rotary phone and expect them to work seamlessly. To make progress, you need a clear plan, smart investments, and the kind of technology that can connect the dots between yesterday's systems and tomorrow's innovations. It's a tricky dance—honoring what works while staying open to the new.

And then there's the never-ending maze of regulatory compliance. As InsurTech keeps pushing boundaries, the rules keep shifting, too. It's not just about staying legal; it's about staying ethical and aligned with what regulators, customers, and even your own teams expect. These two priorities—updating technology and staying compliant—aren't just boxes to check. They're the foundation of how the industry is transforming itself. You can't afford to get either one wrong.

Think of legacy systems like that trusty car you've had for years. It's reliable, and you've added some cool new features over time—maybe a Bluetooth adapter or a fancy GPS. But eventually, there's only so much you can do before it just can't keep up with today's demands. That's why APIs were a game-changer, offering insurers a way to stitch together old and new systems without a total teardown. It worked for a while, but technology moves fast. Those patches won't hold forever. At some point, you're going to need an upgrade—a fully integrated system that's built for the speed and complexity of modern insurance.

And here's the thing about InsurTech: it doesn't wait. By the time you finish this chapter, there will probably be a new startup in the market, and they'll have a shiny new product ready to go in less than a year. That's the pace we're dealing with, and it's why staying ahead means not just playing catch-up but anticipating where the game is headed. The future of insurance belongs to those who can innovate and adapt faster than the market changes—and trust me, it's changing fast.

So, I know that might've added a little to your anxiety, but I had to keep it real and break it down in a way that's simple and clear. My goal is to help you truly understand what you're up against and, more importantly, how you can adapt to it. The evolution of InsurTech isn't about replacing you—it's about bringing in new innovators with fresh perspectives to partner with us and push the industry forward. This is about collaboration, not competition, and if we approach it right, there's space for everyone to thrive.

I talked about the distribution model in the previous chapter, so let's look at the insurtech model as we dive deeper into this chapter. I wanted to lay it out for you and give some examples so that you know what to look for:

Here's a quick breakdown of InsurTechs that are reshaping the distribution side of insurance for agents and brokers. These tools are designed to simplify your workflow, save time, and help you grow your business:

I'll keep saying this: these InsurTech tools aren't here to replace you—they're here to make your job easier and your clients happier. Using the right tools can save time, reduce stress, and grow your business faster than ever.

MGA

Before we go deeper, insurance agents and brokers, let's break down what an MGA—Managing General Agent—is and why it matters to you. Whether you're already familiar with them or just hearing the term for the first time, understanding the role MGAs play in our industry is key, especially as the market evolves with InsurTech.

You will notice that MGAs no longer exist solely for insurance agencies, insurtechs now leverage an MGA to deliver a new insurance product to accompany their technology. This is the #1 delivery model for an insurtech who is creating a brand new product for the market to sell, it allows agents and brokers to get appointed with an MGA almost like they would a carrier. The biggest difference is an MGA is not a carrier, to be a carrier, you have to have a lot more money in the bank, and you have to be willing to take on risks.

So what is an MGA, you ask? At its core, an MGA is a middle layer between insurance carriers and those selling or buying insurance. Unlike traditional agents or brokers, MGAs have special authority granted by the carrier. This authority often includes underwriting policies, setting premiums, binding coverage, and handling claims. In short, MGAs do more than just sell insurance—they have the power to manage the entire insurance process on behalf of the carrier.

For example, imagine a carrier wants to start offering insurance for a niche market, like coverage for small-scale farmers or even cyber liability for small businesses. Instead of building out the expertise and infrastructure themselves, they partner with an MGA that specializes in that market. The MGA handles the underwriting, product design, and operational details, leaving the carrier to focus on capacity and compliance.

For you as an agent or broker, MGAs can be an invaluable partner. They provide access to specialty products you might not get directly from a carrier. Let's say a client approaches you requesting an insurance product outside your usual wheelhouse—maybe it's coverage for a short-term vacation rental or a tech startup's cybersecurity needs. An MGA might be the solution if you don't have a direct carrier relationship for those products. They bring niche expertise and can often respond faster to unique risks.

Now, here's where things get interesting. The MGA model is evolving rapidly thanks to technology. Let's break it down into two types so you can see the difference.

Traditional MGAs are built to serve carriers in specific, often niche, markets. They rely on manual processes and traditional underwriting expertise to deliver specialized products. You've probably worked with these before—they're efficient and effective, but they can also be a little slower and more rigid because they operate within the limits of older systems.

InsurTech MGAs are the modern disruptors of the insurance world. They combine the authority and expertise of a traditional MGA with cutting-edge technology to deliver insurance products faster and with a more customer-focused experience. These MGAs might use AI to streamline underwriting or advanced data analytics to price policies more accurately. The result? Faster turnarounds, more customization, and better tools for you as an agent or broker.

Think of InsurTech MGAs as startups with an insurance license. They're not just offering niche products; they're often creating entirely new ways to approach insurance: from pay-per-mile auto policies to coverage for gig workers to just totally brand new products that solve true consumer needs that we traditionally don't offer coverage for in the insurance industry.

One thing you'll notice about InsurTech MGAs is that their name usually isn't on the paper applications you fill out. They often rely on a reinsurer or fronting carrier to back their products. Many don't have fully in-house claims or underwriting teams, either. Instead, they use advanced technology like AI and machine learning to handle underwriting with oversight, while others have smaller traditional underwriting teams. For claims, they typically partner with TPAs (Third-Party Administrators) initially and then gradually build their own claims teams as they grow.

What I love about the Insurtech MGA model is how it makes creating an insurance product more accessible than ever. With the help of actuarial consultants, reinsurers, fronting carriers, TPAs, and firms like Off Course Consulting and Off Course Tech, almost anyone can build an insurance product from scratch. It's a land of possibilities I never imagined when I started in this industry. At my firm, we help new InsurTech founders,—often with no insurance background—leverage our expertise and connections to determine their distribution strategy and determine if they truly belong in the insurance space or if their idea fits better elsewhere.

My favorite clients are the ones who come to us early, asking what real problems need solving in the industry, instead of those who create something first and then try to force it to fit. Agents and brokers, I know you appreciate it when InsurTechs solve real problems that make your job easier—not just throwing new tech at you that doesn't fit into your workflow. That's the magic of an MGA that gets it right.

So, why should you care about MGAs—whether they're traditional or InsurTech? Because they expand your toolbox. When you partner with an MGA, you're gaining access to new products, markets, and expertise to help you grow your business. InsurTech MGAs, in particular, are pushing the industry forward with new technologies that can streamline processes, improve customer service, and deliver solutions faster.

Here's an example: Imagine working with an InsurTech MGA that lets you quote, bind, and issue policies in minutes—all through a user-friendly app. That kind of speed can help you close more deals and make your clients happier. On the other hand, a traditional MGA might help you land a complex, high-value account by offering the specialized expertise and underwriting capacity you need.

MGAs, both traditional and InsurTech, are reshaping how insurance gets done. They're like your secret weapon for tapping into new markets and delivering solutions that go beyond what a standard carrier relationship might offer. By understanding how they operate and leveraging their strengths, you can better serve your clients and stay ahead in an industry that's evolving faster than ever.

So, whether you're working with a traditional MGA or exploring what an InsurTech MGA can do for you, the key takeaway is this: MGAs are here to make your job easier, your clients happier, and your business more competitive. And in this ever-changing insurance landscape, that's a win-win.

BGA

Now, I do want to clarify the BGA vs. MGA for those who may be a non-P&C agent. A BGA (Brokerage General Agent) and an MGA (Managing General Agent) might sound similar, but they're totally different players in the insurance game. BGAs focus on helping life, health, and annuity agents find the best products, navigate underwriting, and close deals. Think of them as your product plug—they don't have underwriting authority but work with carriers to simplify your job. MGAs, on the other hand, are heavy hitters in property and casualty (P&C) or niche markets, with the authority to underwrite, price, and even manage claims. They're your go-to when you need fast decisions or specialized policies like cyber liability or high-risk commercial auto. So, when does this matter? If you're a life agent, a BGA helps you stay competitive with great products and support. An MGA is your expert partner if you're in P&C or need niche solutions. Both are here to help you serve your clients better and grow your book of business.

Here's a quick chart to help you see the difference:

BGA	MGA
Life, health, annuity focus	Property & casualty, niche focus
No underwriting authority	Full underwriting authority
Supports agents with products	Creates and manages policies
Works only with brokers/agents	Can interact directly with risks

Now that you've explored the roles of MGAs and BGAs, let's take a moment to reflect. Think about the potential partnerships you could build in these areas. Who stands out as a great fit to collaborate with and help grow your agency or brokerage? Consider the types of products or expertise they bring to the table and how they align with your business goals. Jot down a few names or qualities of ideal partners and brainstorm how you can approach them to create meaningful connections.

I must say, I would love for more agents and brokers to become an insurtech founder, you know your problems better than anyone else, so why shouldn't you be the one to solve them? You'll have instant buy-in and trust from your industry peers, and you can also leverage them as focus groups and early adopters for your solution. The industry needs more leaders like you.

If you don't know how to get started, that's what firms like Off Course Consulting are for, we help you go all the way from idea to bringing your vision to reality. We have had the distinct privilege of doing this with a few agency owners, and when I tell you, it's one of the most rewarding feelings I get to experience over and over again. When I encourage someone to do something they never dreamed imaginable and birth something beautiful, it all stems from their ideas right in front of their eyes, it's indescribable. I do get that not everyone wants to be an insurtech founder, and that is okay, I love being able to help them, too. It's been even more rewarding to watch an agency owner save their agency from going under by reinventing their brand and integrating technology, adopting a new business model we created for them, and offering new products they didn't sell before to their clients, a true agency transformation happens right before their eyes.

Let's be honest—many agents and brokers have seen tech and digital transformation efforts fall flat, leaving you wondering if it's even worth the hassle. To change that narrative, we must tackle some common technology myths head-on. These misconceptions can hold you back from embracing tools that can actually make your work easier and your agency or brokerage more successful.

Myth 1: Technology transformation takes too much time.

The truth? It doesn't have to. Tech transformation isn't about changing everything at once—it's about starting small and tackling it step by step. Prioritize the areas that will impact your workflow most and implement one solution at a time. Small wins add up, and before you know it, you'll have transformed your agency without feeling overwhelmed.

Myth 2: Technology doesn't deliver real value.

This couldn't be further from the truth. Agents and brokers who use technology effectively win more business and retain more clients. Why? Because they can deliver faster quotes, provide better service, and create a seamless customer experience. In today's competitive market, clients expect convenience, and tech tools help you meet that demand. They are expecting the "Amazon Effect" for everything these days.

Myth 3: Technology is too expensive for small agencies or brokerages.

Let's break this down: good technology is an investment, not just another expense. Yes, there's a cost upfront, but the return on investment is real. Many tech solutions are scalable, meaning you can start with affordable options that grow as your business grows. Plus, the time you save with automation and efficiency? That's money back in your pocket. Think of it as hiring a 24/7 assistant without the extra payroll. That's what the power of tech like AI or Gen AI can do for an agency or brokerage.

Myth 4: Technology will replace the personal touch clients expect.

Here's the truth: tech doesn't replace your relationships—it enhances them. Clients still want the trusted guidance and expertise you bring to the table, but they also want convenience and speed. With the right tools, you can give them both. Automating repetitive tasks frees you up to focus on what matters most: building stronger, more personal connections with your clients. It's about working smarter, not losing your human touch.

Insurance is a relationship-driven business with agent and broker distribution at the center. When you push past these myths, you'll see that tech isn't the enemy—it's a powerful ally. The key is to focus on technology that solves real problems for your business, not just shiny tools you'll never use. The agents and brokers who embrace it stay ahead, serve their clients better, and grow their books of business faster. Take it one step at a time, and you'll see how the right tech can help you work smarter and retain more clients.

Now, I know what you're probably thinking: "Alexis, how do I even get my agency or brokerage on board with digital transformation?" Listen, I get it—change can feel overwhelming, especially when juggling the day-to-day grind of serving clients. But here's the truth: new technologies are completely transforming our industry, and they're not here to replace what we do—they're here to give us the time and tools to focus on what truly matters: building meaningful relationships with our clients.

The right tools can help you streamline your operations, reduce administrative headaches, and even pull valuable insights from consumer data to level up how you do business. So, let me break it down for you—step by step—on how you can adapt and not just survive, but thrive, in this rapidly evolving landscape. Trust me, if I've learned one thing in my 15 years in insurance, it's that embracing change always pays off. Let's dive in.

Step 1: Cultivate a Digital-First Mindset

1. **Commit to Continuous Learning**: Stay updated on emerging tech trends and tools in the industry. Attend webinars, take courses, or connect with younger agents to exchange knowledge. Training...Education...Training

2. **Shift Your Mindset**: See change as an opportunity, not a threat. The industry is evolving, and adaptability is key to staying relevant and competitive.

3. **Include Your Team**: When implementing new tech, involve your agents and staff to ensure buy-in and avoid wasting resources on tools they won't use.

Step 2: Embrace Technology to Streamline Operations

1. **Automate Repetitive Tasks**: Use batch emailing, automated reminders, and pre-filled forms to save time and reduce admin work. There is definitely an AI tool for that out there, too.

2. **Go Mobile and Paperless**: Adopt mobile-friendly features and digital document storage to make operations smoother and more convenient for your clients.

3. **Consolidate Systems**: Find integrated platforms that combine client management, quoting, and communications in one place to minimize inefficiencies.

Step 3: Enhance the Customer Experience

1. **Meet Clients Where They Are**: Most customers use mobile devices—ensure your services are accessible, fast, and easy to use on the go. You must have a website that is compatible with cell phones.

2. **Offer Choices**: Clients want options. From policy types to payment methods, provide flexibility that mirrors their everyday experiences, like "The Amazon Effect"..

3. **Be Responsive**: Leverage chat tools, email automation, or self-service portals to give clients quick answers and maintain high satisfaction.

Step 4: Collaborate with InsurTechs

1. **Expand Your Product Offerings**: Partner with InsurTechs to introduce innovative products that meet emerging client needs, like cyber insurance, AI insurance, or pay-as-you-go policies.

2. **Tap Into Risk Management Services**: Use InsurTech tools to provide clients with proactive risk management strategies, adding value beyond just selling policies.

3. **Leverage Their Expertise**: InsurTechs bring fresh perspectives and cutting-edge technology to the table. Collaborating with them can modernize your agency without reinventing the wheel.

By following these steps, you'll not only adapt to change but thrive in it, creating a future-ready agency or brokerage that stands out in the marketplace.

Digital transformation doesn't happen overnight, but it starts with intention. Let's make this personal and actionable for your agency or brokerage. Take a moment to reflect on the steps we talked about and jot down your thoughts below. Remember, this is about building something better for you, your team, and your clients.

Step 1: Cultivate a Digital-First Mindset

- **Reflection**: How open are you (and your team) to new technologies? What steps can you take to create a culture that embraces learning and change?

- **Action Item**: Write down one thing you can do this month to encourage collaboration or knowledge-sharing in your agency, like hosting a "tech idea" meeting or asking younger agents for their insights.

....

....

Step 2: Embrace Technology to Streamline Operations

- **Reflection**: Which tasks in your day-to-day feel like they're eating up too much of your time? What processes could be automated or simplified with the right tools?

- **Action Item**: List two areas where you think technology could make a difference—like automating client follow-ups or going paperless. Research one tool that could help and set a goal to test it out.

....

....

Step 3: Enhance the Customer Experience

- **Reflection**: Are you meeting your clients where they are? Do you offer digital options that fit their lifestyle, like mobile access or self-service tools?

- **Action Item**: Think about your own experiences as a consumer. What's one convenience you'd love to offer your clients? Write it down and brainstorm how you can make it happen.

....

....

Step 4: Collaborate with InsurTechs

- **Reflection**: What's one challenge your agency faces that an InsurTech partnership might help solve? Whether it's new product offerings, risk management tools, or operational efficiencies, where do you need support?

- **Action Item**: Write down one InsurTech solution you'd like to explore or a partnership opportunity you think could help grow your book of business. Set a goal to learn more or reach out to a provider this quarter.

....

....

Now, look at what you've written. These reflections and actions are your starting points for transformation. Remember, this isn't about doing everything at once—it's about moving intentionally toward a stronger, more future-ready agency or brokerage. Transformation starts with one step. What's yours?

CHAPTER **SEVEN**

CLOSING THE TALENT GAP IN INSURANCE: STRATEGIES FOR A NEW ERA

Guest Author & Talent Expert: Marvin Vaughn

The topic of the approaching talent gap in insurance has been the talk of the industry for the last few years, so I felt that it would be the perfect transition to bring in an actual talent expert to write this chapter, who has been behind the scenes working on strategies to help close this gap, as the aging generation begins to retire. Marvin Vaughn is a 20+ year Talent Expert, Strategist, and Technology executive who has worked for some pretty incredible companies, including three insurance carriers, one of the world's leading airlines, and a Big 4 Talent consulting firm. Today, Marvin is not just my husband but the COO of Off Course, where he has a talent and operations division that focuses on identifying top talent, creating processes, and developing cultures to retain this top talent alongside top Insurance and Insurtech executives. Introducing the talent guru himself, Marvin Vaughn.

The insurance industry faces a workforce challenge unlike any it has faced before. With seasoned professionals retiring, younger generations looking elsewhere for opportunities, and workplace expectations changing rapidly, it's time for a fresh perspective. This chapter explores the state of recruitment and retention in insurance and how agencies, carriers, and brokerages can evolve to meet the moment.

The insurance industry's talent crisis is real but not insurmountable. At its core, the problem stems from an aging workforce, a lack of appeal to younger generations, and outdated workplace practices. High turnover rates add to the challenge, but the talent gap can be narrowed and eventually closed with the right strategies.

Recruitment is step one. Agencies, carriers, and brokerages must embrace trends like purpose-driven work, diversity, and technology to make insurance careers attractive. Younger job seekers want roles that matter, and showcasing the industry's role in protecting families and businesses can make all the difference. Modern hiring tools and a commitment to inclusivity also ensure that the right candidates are found faster and more effectively.

Retention is where the battle is indeed won. Flexibility, whether through hybrid models or career mobility, is necessary for today's workforce. Employees who see opportunities to grow, learn, and advance are more likely to stay loyal. Continuous learning programs, mentorships, and recognition help create a supportive environment where people can thrive.

The divide between agencies, carriers, and brokerages requires nuanced approaches. Smaller teams thrive on strong community ties and personal development, while larger organizations can leverage their scale to offer diverse career paths. With their fast-paced nature, brokerages should focus on adaptability and technology to retain dynamic talent.

Leadership is another piece of the puzzle. Instead of promoting top sales performers into management by default, agencies need to recognize that leadership is their own skill. Leadership development programs can identify those with the potential to inspire and guide, ensuring that teams have the proper support.

Lastly, technology plays a pivotal role. Outdated systems frustrate employees and hinder productivity. By investing in advanced tools and software, agencies can empower their teams to work smarter, not harder.

Closing the talent gap in insurance requires fresh thinking, bold action, and a genuine commitment to change. By prioritizing flexibility, innovation, and empowerment, the industry can rebuild its workforce and emerge stronger than ever. Together, these strategies will help create a resilient, motivated, and future-ready team capable of meeting whatever challenges lie ahead.

A Sector in Transition

Picture this: you're part of an industry steeped in tradition but struggling to keep up with the speed of change. The insurance sector is at this crossroads, grappling with challenges like:

- **Aging Workforce:** With many seasoned professionals, especially leaders, nearing retirement, there is a pressing need to fill their shoes.

- **Perception Problems:** Let's face it: younger generations don't see insurance as glamorous or innovative.

- **High Turnover:** Employee turnover rates have climbed, fueled by limited growth opportunities and outdated work models.

It's clear that sticking to the same old methods isn't an option anymore.

Recruiting Trends That Matter

Insurance companies need to get creative with recruitment to compete with other industries. Here are some trends to watch—and adopt:

1. **Purpose-Driven Work:** Today's job seekers want to know their work matters. Show them how your agency or company makes a difference, whether protecting families or helping businesses recover after disasters.

2. **Tech-Forward Hiring:** Streamline recruitment with AI and other tools to find the best candidates faster. A tech-savvy process also makes a great first impression.

3. **Career Mobility:** Younger professionals value flexibility and the chance to explore different roles. Highlight opportunities for growth within your organization.

4. **Diversity and Inclusion:** Building a diverse team isn't just good optics—it's essential for innovation and customer connection. Proactively seek candidates from various backgrounds because diverse markets need diverse talent.

Shifting the Culture in Talent Management

Getting talent through the door is just the beginning. Retaining great employees means creating a workplace they won't want to leave. Here's how:

1. **Flexibility is Key:** Whether hybrid work models or flexible hours, today's employees expect their workplace to accommodate their lives.

2. **Learning Never Stops:** Invest in continuous education through certifications, workshops, and mentorship programs. A knowledgeable workforce is a loyal workforce.

3. **Empowered Leadership:** Managers should be mentors, not micromanagers. Provide leadership training to help them inspire and support their teams. Bad leadership can make you lose great talent.

4. **Celebrate Success:** Recognize employees for their contributions—big or small. People who feel appreciated are more likely to stay.

Bridging the Divide Between Agencies, Carriers, and Brokerages

While the whole industry faces similar challenges, each segment needs tailored solutions:

- **Agencies:** Smaller teams thrive on community and close relationships. Focus on clear career paths and robust training programs to keep employees engaged.

- **Carriers:** Large carriers can offer diverse opportunities for lateral movement, letting employees explore different areas without leaving the organization.

- **Brokerages:** Fast-paced and client-facing, brokerages should emphasize innovation and adaptability. Invest in tech tools and cross-functional collaboration.

Recommendations for the Road Ahead

Here are some actionable steps to help close the talent gap:

1. **Master Cross-Selling:** Train employees to identify and meet client needs with complementary products. It's a win-win for your business and customers.

2. **Broaden Skill Sets:** Create job rotation programs and encourage collaboration across departments. This keeps employees engaged and better equipped for future roles.

3. **Rethink Promotions:** Leadership isn't just about sales numbers. Invest in leadership development programs to identify and nurture real leaders.

4. **Upgrade Technology:** Outdated systems frustrate employees. Invest in modern tools that improve efficiency and satisfaction.

5. **Empower Your Team:** Listen to your employees, value their input, and align roles with their strengths. A happy team is a productive team.

The insurance industry is at a pivotal moment. Addressing the talent gap requires a clear understanding of individual strengths and a deep commitment to diversity. Every role within the ecosystem—whether it's the empathetic claims professional, the detail-driven underwriter, or the outgoing agent—demands unique skills that, when properly matched, can transform the workforce into a thriving, innovative force.

Diversity is not just a moral imperative; it's a competitive advantage and plays a critical role in this transformation. The insurance industry's future relies on attracting talent from all backgrounds, experiences, and perspectives, which is essential to building teams that reflect the clients and communities the industry serves. This isn't just about checking boxes; it's about fostering innovation, collaboration, and a workplace where everyone feels valued and empowered. By actively seeking talent from underrepresented groups and creating pathways for all voices to be heard, the insurance sector can move beyond the old ways of thinking and truly evolve.

How do we address the insurance talent gap? By seeing the potential in every person, valuing diversity as a cornerstone of progress, and ensuring that the right people are in the right roles.

CHAPTER **EIGHT**
THE WISDOM CHAPTER

Welcome to the Wisdom chapter. I chose this as the final chapter because it weaves together everything this book has taught you and brings it to life in human form. As you turn these last few pages, I want you to envision yourself thriving in a long, successful insurance career—seeing pieces of your own journey in the stories and lessons shared. I've called on some of the best and brightest in the industry, and they answered the call, to share their stories and wisdom with you.

The Wisdom of Bonnie Boone: Lessons from a Trailblazing Career

Name: **The Iconic Bonnita (Bonnie) Boone**

Years in the industry: **45 years**

Agent or Broker or MGA: **Broker**

Designation: **None**

Company: **First African American Broker and Former Executive of multiple Big 4 Firms**

When I asked Bonnie Boone about the most important lesson she's learned about building and maintaining client relationships, her response was as straightforward as it was powerful: "Responsiveness and foresight are everything,". "You can't just focus on today's needs, like the auto or D&O policy on your desk. You've got to anticipate what's coming down the road and help your clients prepare for emerging risks." Bonnie's ability to think critically and act proactively has set her apart throughout her career, ensuring her clients saw her as both a service provider and a strategic partner.

Bonnie's trailblazing career is full of incredible accomplishments. She was the first Black broker to work for a Big Four firm, and she trained three men who are now presidents of major insurance companies—leaders at Liberty Mutual, Transatlantic Re, and ProAccess. "I've always believed in mentoring and building others up," she explained. "Training those men wasn't just about teaching them the technical side of insurance; it was about showing them how to think critically, act ethically, and lead with empathy." Her work didn't just shape individual careers; it transformed the culture of organizations.

When I asked Bonnie how the role of agents and brokers has evolved since she started, she reflected on how much the industry has changed. "When I joined Marsh, there were only 10 Black professionals out of 49,000 employees," she recalled. "Back then, I thought professionalism meant adapting to their culture—wearing my hair back, dressing a certain way, and keeping my head down. Women couldn't even wear slacks or open-toe shoes in the office." Today, while the industry has made progress, Bonnie sees new challenges for younger professionals. "I always tell them to pay attention to the culture of the firm they join. If it doesn't align with your values, it'll be hard to thrive there."

Bonnie's career has been defined by her ability to think creatively and innovate. "One of my proudest moments was when I introduced cyber liability into medical malpractice coverage," she said. "I realized that if someone entered the wrong data in a hospital system and caused a medical error, the liability wasn't just cyber or malpractice—it was both." That kind of forward-thinking approach not only earned her respect in the industry but also shaped policies that better protect clients. Her creativity extended beyond traditional insurance roles, including her work insuring the Biden-Harris campaign, a responsibility that required both technical expertise and immense trust.

When I asked Bonnie what advice she would give to someone starting out in the industry, she emphasized two things: be strategic and become an expert. "Early in my career, I thought technical competence would be enough. I could tell you every exclusion in a policy and why something wasn't covered, but I wasn't always as strategic as I should have been," she admitted. "You need to know when to speak up and when to hold back. Also you've got to position yourself as an expert—write articles, speak at conferences, and make sure people know you're the go-to person in your field."

Finally, I asked Bonnie what she believed to be the biggest challenge facing the insurance industry today. She reflected on the younger generation of professionals entering the field. "They're incredibly smart, but I think many lack the creativity to think beyond what they've been taught," she said. "It's not enough to follow a formula. You've got to think past the norm and figure out innovative solutions for your clients." At the same time, Bonnie acknowledged how much younger professionals have taught her about balance. "When I started, it was all about work, but now I see the importance of life-work balance. It's something I've come to appreciate later in life, and it's a lesson I carry with me every day."

Bonnie Boone's illustrious 45-year career in insurance is a masterclass in leadership, resilience, and trailblazing success From breaking barriers as the first Black broker at a Big Four firm to shaping the next generation of insurance leaders, she has set a standard for excellence that continues to inspire. Her accolades alone deserve a round of applause. Recognized as a **Power Broker** by *Risk and Insurance Magazine* in 2008 and 2009, she set the standard for excellence early in her career. Bonnie's influence grew, earning her a spot among *Business Insurance*'s prestigious **Women to Watch** in 2008 and 2015. Her dedication to diversity and inclusion shone brightly with her nomination for the **IICF Inclusion Champions Award** in 2017. That same year, she was celebrated with the **Elite Women Award** by *Insurance Business America*, an honor she reclaimed in

2021. By 2020, her legacy was cemented when she was inducted into the **Insurance Business America's Insurance Hall of Fame**. As if that weren't enough, she received the **Lifetime Achievement Award** from the National African American Insurance Association (NAAIA) in 2019, further solidifying her role as a pioneer. *Savoy Magazine* named her one of the **Most Influential Women in Corporate America** that same year. In 2021, she was recognized by the National Diversity Council as one of the **Most Influential African American Leaders in Business**. Bonnie Boone's accolades are not just a testament to her professional achievements—they reflect her unwavering commitment to lifting others as she climbed. Her story is nothing short of legendary.

Wisdom from Lisa Lounsbury: Building Relationships and Thriving in Insurance

Name: **Lisa Lounsbury**

Years in the industry: **29 years**

Agent or Broker or MGA: **Independent Agent Advocate and Former Insurance Carrier**

Company: **Big I New York-President**

When I asked Lisa Lounsbury, president of Big I NY, about the most important lesson she's learned in building and maintaining strong client relationships, her response was refreshingly simple yet profound: "These are human relationships,". "It's not just about selling; it's about caring—caring about your clients, their families, and their goals. If you invest time into really knowing them, listening to their needs, and being there for them, you'll become more than just an agent. You'll be their trusted advisor." Lisa emphasized the power of small gestures—sending a thoughtful handwritten note or checking in on a client's milestone—and how these moments can strengthen the bond between agents and their clients.

I then asked her how the role of agents and brokers has evolved over the years. Lisa broke it down by generation, explaining that for more seasoned agency owners, the critical gap is technology.

"Many agency owners in their 50s and 60s excel in technical insurance knowledge but haven't fully embraced technology. They need to invest in people who can implement tech solutions that elevate the customer experience," she explained. She offered a different perspective for younger agents: "The new generation is naturally tech-savvy, but they don't always understand the importance of mastering insurance fundamentals. The winners in this industry will be those who balance their tech skills with deep product knowledge. At the end of the day, insurance is still about ensuring clients have the right coverage."

The conversation shifted to the challenges of balancing technology with human relationships. Lisa stressed that technology should enhance, not replace, the personal connections that define the industry. "Too often, agents chase the shiny new tech without understanding its ROI or how it fits into their operations. The most successful agents will adopt technology strategically—to support relationships and improve retention—while keeping the human touch at the forefront," she explained. Lisa highlighted the enduring value of relationships, especially when working with carriers. "Strong relationships with underwriters and carriers will always be key even in a digital age. Tech can't replace trust."

When I asked Lisa about the biggest challenges facing the insurance industry today, she pointed to the regulatory landscape and the need for ethical use of data. "The patchwork of state regulations complicates everything—from products to pricing," she said. "And as technology advances, agents will need access to data to serve clients better, but they must use it ethically." Lisa also noted how legal system abuse and climate change are reshaping risk management, adding, "Agents can't just sell policies anymore; they need to help clients prevent losses and think more strategically about risk."

I wanted her perspective on advice for newcomers to the industry, and her response was clear: "Get a mentor—or several," Lisa urged. "Whether it's an experienced agent, a carrier representative, or someone in a different role, mentorship is invaluable. This is an industry where people want to help the next generation succeed.

Build those relationships, ask questions, and learn from seasoned professionals." She shared a heartwarming story of pairing a seasoned agent with a new mentee and how their knowledge exchange benefited both.

Finally, I asked her about the one piece of wisdom she wished she had known earlier in her career. "It's a small world," Lisa said. "This industry is a tightly connected family, and the relationships you build today will matter decades down the road. I've stayed in touch with colleagues I met in the '90s, and those connections have opened doors I never expected." Her advice? "Don't burn bridges. Be generous, work hard, and always go the extra mile. That's how you build a career that lasts."

Lisa's insights offer a powerful reminder that insurance is a people-first business. It's about mastering the fundamentals, leveraging technology wisely, and building relationships that stand the test of time.

Wisdom from Tinsley English: Lessons from 25 Years in Insurance

Name: **Tinsley English**

Years in the industry: **25 years**

Agent or Broker or MGA: Broker: **Broker**

Designation: **CIC, CRM, CDP**

Company: **Big 4 Firm Brokerage and Author of** *Grit, Growth and Gumption for Women: Three Keys to Lead Yourself and Others with Confidence*

Tinsley English, a seasoned operations executive, author of *Grit, Growth, and Gumption*, and leader with 25 years in the insurance industry, brought fresh perspectives when I asked her about the lessons she's learned in her career. Starting with client relationships, Tinsley emphasized the power of clear, responsive communication. "If you can't meet a deadline, pick up the phone and let your client know. Be honest and transparent," she said. For her, it's not just about efficiency but also about building trust. "Get to know your clients as people—their kids, their pets, even their favorite Starbucks drink. These small touches create a bond that goes beyond business." She also highlighted the value of maintaining strong relationships with colleagues, noting that these connections often lead to unexpected career opportunities.

When discussing professional growth, Tinsley underscored the importance of designations like the CRM and CIC. "These certifications put you in the buyer's seat, helping you understand the priorities of CFOs, HR leaders, or safety managers. Knowing your audience is key," she explained. She encouraged newer professionals to take advantage of these opportunities early in their careers. "If your company offers to pay for designations, go for it while you have fewer personal commitments." For seasoned professionals, she suggested programs that combine designations, such as pairing CPCU with an ARM, to stay competitive and deepen expertise.

On the industry's evolution, Tinsley reflected on the advantages and pitfalls of technology. "We can work faster and more efficiently now, but the art of picking up the phone is getting lost," she lamented. "Don't hide behind emails—pick up the phone and talk. This skill builds trust and shows clients that you care." She also stressed the importance of having a servant's heart. "You're in the service industry. Clients and underwriters alike should feel that you have their best interests at heart. It's a give-and-take relationship."

When I asked her about the biggest challenges in the industry, Tinsley pointed to leadership and emotional intelligence. "Everyone is a leader, no matter their title," she said. She advocated for emerging leadership programs to teach resilience, empathy, and emotional intelligence. "This is a tough industry—stressful and fast-paced. People need to learn how to lead themselves first before they can lead others." She also highlighted the importance of understanding team dynamics through tools like personality assessments to maximize strengths and develop weaker areas.

Tinsley was clear on balancing human expertise with technology in a more digital future: "AI is here to stay, but we must layer it with human reasoning to maintain integrity." She sees the future as a collaboration between tech and humanity, with technology serving as a tool rather than a replacement for expertise.

Finally, I asked Tinsley for the one piece of wisdom she wished she'd known earlier. Her response was bold: "Color outside the lines. Don't limit yourself to the four corners of your job description. Rewrite it. Innovate. Create a new role if you see a gap." For Tinsley, success comes from working within the system and finding ways to improve it. This advice, like her career, is a testament to her grit, growth, and gumption.

Wisdom from Diana Greenberg: Confidence, Change, and the Art of Adaptation

Name: **Diana Greenberg**

Years in the industry: **39 years**

Agent or Broker or MGA: **Agent and BGA President**

Designation: **No**

Company: **Simplicity Group-Partner**

Diana Greenberg, a partner at Simplicity Group and president of Total Financial, shared her insights on building a career in the evolving insurance industry. With nearly four decades of experience, she reflected on how the landscape has shifted dramatically since she started in 1985. Diana explained that in the early days, life insurance was about one thing: protection. Agents knocked on doors or made cold calls from the phone book, selling policies to ensure families were financially secure in the event of a loss.

Today, she said, insurance has evolved into a versatile financial tool used for wealth building, income generation, and living benefits. "It's now considered an asset," she said. "The role of the agent has grown to include conversations about wealth management and investment, making it essential to understand how insurance fits into a client's overall financial picture."

When I asked her what skills are most critical for professionals entering the field today, Diana was adamant about the importance of product knowledge. "You can't sell what you don't understand," she stressed. Knowing the nuances of various policies—whether they offer term protection, cash accumulation, or supplemental income—allows agents to match the right product to the client's specific needs. Equally important, she said, is the ability to communicate effectively. "You have to start by understanding what the client wants to achieve, not by offering a solution right away." Diana also emphasized that adapting to technology is no longer optional. "If you're unfamiliar with AI, digital platforms, and data analytics, you'll fall behind," she said.

Diana pointed to education and technology as the biggest challenges facing insurance distribution today. "Education has to come first," she said. "We need to train new agents on the fundamentals and how to use technology to simplify the process for clients." She noted that technology, while crucial, hasn't yet replaced the need for agents, especially for complex cases. However, she foresees a future where smaller, simpler policies might be sold entirely online, with platforms like Amazon potentially stepping into the space. "For larger, more nuanced policies, agents will still be critical. There's no substitute for human expertise in explaining the complexities of a policy," she explained.

When reflecting on what makes client relationships successful, Diana emphasized transparency, trust, and keeping your word. "I'm known for being straightforward, and I don't make promises I can't keep," she said. She believes in being a true partner to both clients and carriers, ensuring that every transaction benefits all parties involved. "You can't just push a policy to make a sale; it has to be the right fit for the client and sustainable for the carrier," she explained. Her approach—honest, data-driven, and client-focused—has earned her respect across the industry.

For those starting out in insurance, Diana advised seeking mentors and aligning with established professionals or organizations. "You need to learn from those who've been in the field and understand its history," she said. She also suggested that while technology has reduced the need for large office staff, mentorship and partnership remain invaluable. "Work with someone who has the wisdom and experience you can't get from a textbook," she added.

Finally, I asked Diana what single piece of wisdom she wished she'd known earlier in her career. Her answer was simple but powerful: confidence. "You have to believe in yourself and your expertise," she said. "It's okay not to know something right away but be confident enough to find the answer and follow up. Clients trust confidence, and it's what will set you apart." Her journey, marked by adaptability, honesty, and a relentless commitment to learning, offers a masterclass in how to thrive in an industry that's constantly evolving.

White Glove Wisdom: Insights from Marcus Greene

Name: **Marcus Greene**

Years in the industry: **17 years**

Agent or Broker or MGA: **Agent, FMO CEO, and Senior Market Expert**

Designation: **MBA**

Company: **HealthPro Consultants**

In Marcus Greene's 17 years in the insurance industry, he's learned the art of standing out in a crowded market. "We all sell the same commodity," so the key is what I call the white glove treatment." For Marcus, this means going beyond the expected: handwritten birthday cards, personal phone calls, and building a family-oriented atmosphere where clients feel special and valued. Relationships aren't limited to clients either. "Among peers, reputation is everything," Marcus emphasized. "You have to be cordial, even with competitors. You never know—you might end up working together or they may acquire your practice. Protecting your reputation is paramount."

Reflecting on the evolution of the insurance space, Marcus highlighted how technology has reshaped client acquisition. "When I started, it was all cold calling from a Dun & Bradstreet list," he laughed. "Today, the digital age makes reaching decision-makers more challenging." This shift has driven some agents out of the industry, especially those hesitant to embrace new tools or hire experts to guide them. As for skills, Marcus stressed that the fundamentals—like determination and customer service—remain vital, but agents must now adopt a consultant mindset. "The agent's role has shifted from being transactional to being a trusted advisor. Clients come to us informed; our job is to deepen the conversation and help them strategize."

With the influx of AI and automation, Marcus believes agents must pivot to roles that technology can't replicate. "Bots can handle basic enrollments, but they can't offer empathy," he explained. "Agents have to position themselves as problem solvers and long-term strategists. For example, clients want to know: Will my insurance work when I need it? Will I go broke if something catastrophic happens? Is this within my budget? Those are nuanced conversations that require human understanding." Embracing AI as a co-pilot rather than fearing it as a competitor is the key to staying relevant.

Marcus pointed to the challenge of navigating artificial intelligence without regulation as a double-edged sword. "AI has incredible potential, but it also 'hallucinates'—producing inaccurate results that could lead to lawsuits," he said. Agents should focus on how technology can enhance their practices rather than replace them. "Leverage AI to streamline processes, but keep the human element intact," he advised. The balance of tech and touch will determine the industry's trajectory.

For those just starting, Marcus emphasized the importance of focus and expertise. "Pick a niche—whether it's Medicare, P&C, or life insuranceand give yourself at least five years to build a strong book of business," he said. He cautioned against the allure of chasing multiple streams of income too early. "Focus on one thing and do it well," Marcus urged. "Look at LeBron James or Steve Jobs—they mastered their craft before diversifying. There is power in expertise. Build something sustainable first."

Reflecting on his own career, Marcus admitted he wished he had embraced venture capital earlier to scale faster. "I should have gone down that road sooner," he said, noting the opportunities for growth in an evolving industry. Above all, his journey has taught him the value of being adaptable and forward-thinking. "This is a long game," Marcus concluded. "Invest in yourself, your relationships, and your expertise, and the rest will follow."

Wisdom from Jeanette Wilson: Underwriting Insights

Name: **Jeanette Wilson**

Years in the industry: **23**

Agent or Broker or MGA: **MGA and Carrier Background**

Designation: **CIC, TRIP, STAR**

Company: **Insurance Carrier-Underwriting Executive**

In my conversation with Jeanette Wilson, a seasoned underwriting executive with over two decades of experience across carriers and MGAs, I unearthed invaluable insights into the world of insurance. Jeanette's advice not only offered a glimpse into her career journey but also illuminated pathways for others navigating this dynamic industry. Here's what she shared, shaped by years of wisdom and deep expertise.

When asked about the most important lesson she's learned in building strong client relationships, Jeanette's answer was simple but powerful: *pick up the phone*. "This is and will always be a relationship business," . Honesty, follow-through, and genuine care for the client's success are non-negotiables. Jeanette emphasized the importance of listening and being transparent, qualities that make clients feel confident in your dedication to their needs. Relationships, she explained, are the backbone of long-term success.

Reflecting on the evolution of the industry, Jeanette noted how agents and brokers have transformed from transactional sales-people to trusted partners. Today, the focus is on guiding clients through risks and providing customized solutions. For the next generation, she highlighted the importance of emotional intelligence, deep product knowledge, and technical savvy. "Being able to connect personally is just as crucial as mastering the technical side," she said, underscoring the human element of this business.

For those just starting out, Jeanette's advice was both practical and reassuring: *be curious, flexible, and willing to learn.* Shadowing experienced colleagues, pursuing certifications like CPCU or ARM, and building a strong network can provide a solid foundation. Mistakes are inevitable, but Jeanette encouraged embracing them as learning opportunities. "This industry isn't fragile—you're not going to break anything. What matters is how you learn and grow from each experience," she added, stressing the importance of patience and consistency.

On the challenges facing the industry, Jeanette identified technological disruption and emerging risks like cyber threats as major hurdles. However, she sees these challenges as opportunities for those who stay adaptable. "Learn about trends like ESG, AI, and climate-related risks—these are shaping the future of insurance," she advised. Professionals who lean into change and seek to understand the bigger picture will not only survive but thrive in this shifting landscape.

With technology transforming the way insurance is sold and serviced, Jeanette was clear that while automation is efficient, it can't replace the human touch. Agents and brokers are increasingly valuable as advisors who can simplify complex issues and guide clients through tough decisions. "The key," she noted, "is knowing when to use technology and when to step in personally. Striking that balance is what sets great professionals apart."

Finally, when asked about a piece of wisdom she wished she'd known earlier, Jeanette shared a reflective truth: "Focus on relationships." Early in her career, she admitted to being too target-driven, which sometimes led her to overlook opportunities to build lasting connections. Over time, she learned that clients who trust you bring not only loyalty but bigger opportunities. Her advice? "Invest in people, not just transactions."

Jeanette's career is a testament to the power of adaptability, emotional intelligence, and relentless curiosity. For anyone stepping into or advancing within the insurance world, her insights serve as a roadmap to success—and a reminder of the value of building trust, both with clients and within the industry.

Wisdom from Taffy Jo Mayers: The Crooked Path to Innovation

Name: **Taffy Jo Mayers**

Years in the industry: **30 years**

Title: **Chief Commercial Officer-Digital Distribution**

Agent or Broker or MGA: **Brokerage-Underwriter and Head of Distribution**

Designation: **No**

Background: **Carrier and Brokerage**

Company: **Big 4 Brokerage Firm**

When I asked Taffy Jo Mayers about her journey in insurance, her candor and insight struck me immediately. With over 30 years in the industry, she's navigated it all, from early-career pivots to spearheading digital transformation. Her story is one of adaptability, resilience, and a firm belief in the power of innovation.

When I asked Taffy about how she ended up in insurance, she said, "I was going to change the world through public policy," recalling her early aspirations. But reality led her elsewhere. "I needed to make ends meet, so I took a risk management class. Th at class changed my life." What started as a detour became a calling. She stressed the value of lexibility: "Don't take the straight path. The crooked path will teach you more and get you where you truly need to be."

Taffy Jo is at the forefront of digital transformation in the insurance space, and her perspective on technology is both practical and visionary. "We've been discussing digital transformation for decades," she said. "But we've spent too much time automating the old coffeehouse instead of reimagining it." She explained that the industry's core principles—risk pooling, large numbers, and Bernoulli's theories are timeless, but how we operate must evolve. "Technology isn't about replacing people. It's about giving us the tools to make faster, smarter decisions."

She described the role of brokers and underwriters in this digital era: "Brokers need to be solution architects, and underwriters should think like data scientists. You can't just be a salesperson or a risk calculator anymore. You have to combine expertise with technology to deliver real solutions."

Taffy's heart for public policy never left her, which shows in her work. "Insurance is about building resilient communities," she said. One of her proudest moments was working on Chicago's Plan for Economic Growth and Jobs, a public-private partnership that used risk management and technology to create smarter, safer neighborhoods. "The work reminded me why I love this industry. We have the power to make a real difference when we focus on our public mission."

When I asked about the biggest challenges facing the industry, Taffy didn't hold back. "We've lost the public's trust," she said. "We're so risk-averse that we've created coverage deserts, leaving entire communities and industries unprotected." She believes the key to rebuilding that trust lies in collaboration and transparency. "We need to use technology to share data and insights,not just for profitability, but to make the world safer and more equitable."

Taffy also highlighted the role of regulators. "We've treated regulators like the big bad wolf for too long. Instead, we need to bring them into the conversation. As we've seen in fintech, collaboration with regulators can drive real innovation."

Taffy's advice for those entering the field was direct: "Ask questions and listen. This industry demands a hunger for learning." She emphasized the importance of understanding the fundamentals—policy wording, pricing structures, and risk assessment. "Technology is a tool, but if you don't understand the basics, you can't use it effectively."

As our conversation wound down, I asked Taffy for one piece of wisdom she'd impart to those of us still carving our paths. She smiled and said, "Embrace the gray." She explained that success isn't about sticking to a rigid plan or seeing the world in black and white. "The zigzags in your career teach you the flexibility you need to thrive," she said. "Keep your spine strong, but don't let it become stiff." In an industry and a world that's constantly evolving, adaptability isn't just an asset; it's a necessity.

Leaving our meeting, I felt a renewed sense of purpose. Taffy's insights on technology, her commitment to societal betterment, and her emphasis on adaptability resonated deeply with me. In a field that often resists change, her vision offers a roadmap for how we can all move forward by embracing the gray areas, leveraging technology for good, and never losing sight of why we entered this industry in the first place.

Wisdom from James Wong: The Collaborative Edge

Name: **James Wong**

Years in the industry: **26 years**

Agent or Broker or MGA: **Entrepreneur & Wholesaler**

Designation: **MBA**

Company: **The Founder's Chair**

When I asked James Wong what inspired him to enter the insurance industry, his answer was refreshingly direct. "This wasn't always going to be my path," he admitted. For James, insurance became a way to combine his entrepreneurial spirit with a sustainable business model that solved real problems. He entered the field at a transformative time, witnessing the shift from old-school door-knocking and cold calls to internet-based lead generation. "I saw an industry ripe for change, and I wanted to be part of that transformation," he shared. Early on, James realized the power of technology to revolutionize how agents connect with clients and deliver solutions.

When we discussed the future of the industry, James emphasized the enduring importance of soft skills. "Technology will evolve, but people still want connection," he said. "Be personable, be likable, and make conversations organic." He explained that while digital tools can enhance efficiency, they can't replace the human element. "AI won't ask about your kids or your family vacations," he said. For James, the ability to connect on a personal level is what sets great professionals apart, even in a tech-driven world.

James reflected on a critical lesson he's learned about maintaining trust with clients and peers. "Be upfront about your value," he advised. "Tell people how you'll serve them and how you're compensated—don't shy away from hard conversations." He highlighted the importance of clear communication to avoid eroding trust. "When expectations aren't aligned, problems arise," he explained. For James, a handshake should be backed by clarity and mutual understanding, setting the stage for long-term relationships.

For those entering the industry, James stressed the importance of passion and networking. "You have to like people or at least enjoy solving their problems," he said. He encouraged new professionals to leverage social media to build a presence and connect with others in the field. "Talk to people, ask questions, and find your niche," he suggested. James believes that by engaging with mentors and exploring various roles, newcomers can discover paths beyond traditional sales, whether in marketing, technology, or other facets of the industry.

When I asked about the biggest challenge facing insurance today, James didn't hesitate. "The industry has a reputation problem," he said, referencing public perceptions around commissions and junk fees. He advised agents and brokers to demonstrate their value to carriers and clients by creating efficient, client-centered practices. "We need to prove that independent distribution works," he said. By showing creativity and efficiency, professionals can highlight the benefits of their approach and push the industry toward positive change.

Finally, I asked James for one piece of wisdom he wished he'd known earlier. "This industry is highly collaborative if you find the right people," he said. Reflecting on his own struggles, James admitted that he once believed success was only for top salespeople. "If someone had told me earlier that there are many paths to success in insurance, it would've saved me a lot of frustration," he shared. Today, James encourages others to seek out supportive mentors and authentic connections. "The industry is full of people willing to share what they've learned. Don't give up—just find your tribe."

James Wong's journey is a testament to adaptability, authenticity, and the power of human connection in an industry that's constantly evolving. His insights remind us that success isn't just about mastering products or processes, it's about building trust, embracing change, and staying true to your purpose.

Wisdom from Jeff Shi: Building Success Without Sacrificing What Matters

Name: **Jeff Shi**

Years in the industry: **14 years**

Agent or Broker or MGA: **Agent**

Designation: **None**

Company: **Insurtechs.Io-Chief Storyteller**

When I asked Jeff Shi about building strong client relationships, his response was simple yet profound: "We're in a service business." He emphasized the value of human connection, especially in a world where many companies rely on automated systems. "Sometimes customers aren't paying for a cheaper price; they're paying to talk to a human being," he said. For Jeff, the key to setting your agency apart is making it easy for clients to reach someone who will listen. "People just want to feel heard—whether they're venting, asking questions, or seeking solutions."

Jeff has seen the insurance industry shift dramatically since he started in 2011. At that time, CRM systems and internet leads were gaining traction but viewed by many as fleeting trends. "Now, companies have built billion-dollar public businesses on internet leads," he pointed out. Today, the landscape continues to evolve with AI and digital wallets taking center stage. Jeff warned, "We're on the doorstep of agents and brokers competing with AI, so you need to have tools in your office to stay relevant."

When I asked Jeff what skills are critical for agents to succeed, he highlighted trust, technology, and adaptability. "The word that should always be next to an agent's name is trust," he said. "If you lose that, you lose your core value." He advised professionals to stay informed and leverage AI tools to enhance their services. But above all, "Know your stuff, know your products, and always answer clients' questions with truth."

Jeff's insurance journey began with a single retail agency opened with $140,000 in savings. Despite being turned down by major carriers, he grew his first agency to $4 million in just two years. Over the years, he scaled to multiple agencies, launched a franchise model with multiple locations, and explored ventures in recruiting and media. Reflecting on his success, Jeff shared a Chinese saying: "A students work for C students." His point? Risk-takers often reap the greatest rewards. "If two lions are in the same field, the one willing to take risks will be the one eating."

One of the most striking pieces of wisdom Jeff shared was about balance. "Don't sacrifice your family," he told me. "We only succeed 94% of the time. You can burn the boat, but don't burn your family." For Jeff, success is not worth the cost of missing life's most meaningful connections.

As our conversation closed, Jeff offered advice to new professionals and seasoned founders alike. "This is a long journey. Smile," he said. He urged aspiring agents to ask questions, embrace challenges, and stay curious. For founders, his message was clear: "Take risks, but don't lose sight of what matters most—your family and your integrity." Jeff's wisdom reminds us that success is more than profits; it's about building trust, nurturing relationships, and finding joy.

Wisdom from Rex Hickling: The Journey to a Billion

Name: **Rex Hickling**

Years in the industry: **40 years**

Agent or Broker or MGA: **Aggregator/Network**

Designation: **CPCU**

Company & Title: **Premier Group Insurance-President**

When I asked Rex Hickling how he built Premier Insurance Group into a billion-dollar agency, his answer was both simple and powerful: mindset and culture. Drawing on 25 years of carrier-side experience, Rex embraced a philosophy he called the "Three I's: constant improvement, investment, and innovation." Paired with the "Three Q's: quick, quantity, and quality," this framework set the foundation for transformation. "We don't just aim to be better," he told me. "We aim to be the best." This approach drove Premier to hit $100 million in premium in 2015, followed by $1 billion in 2025. Their next goal? $2 billion by 2030—a target Rex views as ambitious yet attainable, fueled by consistent focus and strategic growth.

Rex emphasized the value of learning from the captive insurance model while bringing its strengths to the independent space. "Captives have incredible branding and structured processes," he said. "When you bring that discipline into the independent world, coupled with the freedom of choice, you can truly go national." By committing to excellence and fostering a culture of collaboration, Premier positioned itself as a leader in the independent agency channel.

When I asked how agents could embrace the digital future, Rex offered practical advice. "Start with your carriers," he said. "Insurance companies often have digital programs, tools, and branding resources you can use for free. Leverage those—they're betting on you." He also stressed the importance of balancing technology with human connection. "Automation should free you to focus on relationships, not replace them. Always give clients an option to reach a human when needed."

Rex's approach to growth is rooted in accountability and collaboration. He recalled challenging Premier's founder to commit to being the best before joining the agency. Together, they rebuilt the organization with an external focus, ensuring clients, agents, and carriers succeeded. "When you help others succeed, your own success follows," he explained. Their steady growth was driven by a relentless focus on improving operations and aligning their goals with their stakeholders' needs.

"Listening is where it all begins," Rex shared when I asked about building strong client relationships. He emphasized finding common ground, understanding needs, and building trust. But listening is just the start. Rex's approach to selling insurance is rooted in storytelling. "Don't give a technical definition; paint a picture," he advised. He shared an example of explaining uninsured motorist coverage by connecting it to a client's child. "People don't buy insurance; they buy peace of mind."

Reflecting on industry challenges, Rex highlighted the need for agencies to adapt to change. "Work *on* your business, not just *in* it," he said. This means evaluating workflows, automating tasks, and freeing up human capital for meaningful interactions. Rex also pointed out the importance of empathy in a hard market, where rising premiums strain both clients and staff. "Support your team and show them you're in this together," he advised. "That confidence builds resilience."

When I asked Rex what wisdom he wished he'd known earlier, his answer was deeply personal: "It's not about being right—it's about connection." Early in his career, he learned the value of listening and collaboration over asserting dominance. He also stressed the importance of lifelong learning. "Sharpen your saw," he said, quoting Stephen Covey. Whether through books, carrier councils, or mastermind groups, Rex believes in continuously improving. With his focus on collaboration and growth, Rex has demonstrated how listening, learning, and leading can drive exceptional results—and inspire others along the way.

I hope you gained valuable wisdom from my industry friends, sheros, and heroes. It has been an absolute honor to walk alongside you, sharing my passion for this incredible industry and the vital role agents and brokers play in it. Remember to keep learning, stay curious, and never give up. The road may be winding, but it's worth every step. Here's to your journey, your growth, and the legacy you'll leave in this ever-evolving world of insurance distribution. The future is now! Until we meet again in the next volume...**~Alexis Cierra Vaughn~**

www.ingramcontent.com/pod-product-compliance
Lightning Source LLC
Chambersburg PA
CBHW040859210326
41597CB00029B/4907